Software Engineering Requirements Knowledge Areas

Volume 1:

The Engineering of Software Projects

The Engineering of Software Projects
Software Engineering Knowledge Areas

These twelve volumes support the IEEE *Guide to Software Engineering Body of Knowledge (SWEBOK)* and the IEEE *Computer Society Professional Software Engineering Master Certification* exam.

This is a work-in-progress; each of these volumes is not currently available, but is scheduled to be published in the coming year.

—Richard Hall Thayer

Volume 1 — Software Engineering Requirements

Volume 2 — Software Engineering Design

Volume 3 — Software Engineering Construction

Volume 4 — Software Engineering Testing

Volume 5 — Software Engineering Maintenance

Volume 6 — Software Engineering Configuration Management

Volume 7 — Software Engineering Management

Volume 8 — Software Engineering Processes

Volume 9 — Software Engineering Models and Methods

Volume 10 — Software Engineering Quality Assurances

Volume 11 — Software Engineering Economics

Volume 12 — Software Engineering Project Management

These volumes will be published and sold through Amazon Books.

Software Engineering Requirement Knowledge Areas

Volume 1:

The Engineering of Software Projects

Richard Hall Thayer, PhD, CSDP

Contributing authors:

Peter H. Sawyer, Professor, Lancaster University, U.K.
James D. Palmer, PhD, George Mason University

Software Management Training
Carmichael, California
2017

Table of Contents
Software Engineering Requirements Knowledge Areas

A Partial List of General Abbreviations
(One-of-a-kind abbreviations are identified in place)

a.k.a.	—	also known as
a.s.a.p.	—	as soon as possible
ConOps	—	concept of operations (document)
CSCP	—	Computer Society Certificates of Proficiency
DFD	—	data flow diagram
GUI	—	graphical user interface
HCI	—	human computer interface
HW	—	hardware
I/O	—	input/output
IDE	—	integrated development environment
IV&V	—	independent verification and validation
KA	—	knowledge area
PSEM	—	Professional Software Engineering Master (Certification)
SCM	—	software configuration management
SED	—	software engineering design
SEM	—	software engineering management
SEPM	—	software engineering project management
SET	—	software engineering testing
SQA	—	software quality assurance
SW	—	software
SER	—	Software engineering requirements
SWE	—	software engineering
SWEBOK	—	(Guide to the) Software Engineering Body of Knowledge
TBD	—	to be determined/done
V&V	—	verification and validation

Honorary Foreword

To explain the origin of the term "software engineering," the following story is offered.[1]

In the mid-1960s, there was increasing concern in scientific quarters of the Western world that the tempestuous development of computer hardware was not matched by appropriate progress in software development. The software situation looked to be more turbulent. Operating systems had just become the latest rage, but they showed unexpected weaknesses. The uneasiness had been articulated in the NATO Science Committee by its U.S. representative, Dr. I.I. Rabi, the Nobel laureate and famous, as well as influential, physicist. In 1967, the Science Committee set up the Study Group on Computer Science, with members from several countries, to analyze the situation.

The German authorities nominated me for this team. The study group was given the task of "assessing the entire field of computer science," with particular elaboration on the Science Committee's consideration of "organizing a conference and, perhaps, later on setting up an "International Institute of Computer Science."

The study group, concentrating its deliberations on actions that would merit an international rather than a national effort, discussed all sorts of promising scientific projects. However, it was rather inconclusive on the relation of these themes to the critical observations mentioned above, which had guided the Science Committee in creating the study group.

Perhaps not all members of the study group had been properly informed about the rationale for its existence. In a sudden mood of anger, I remarked, "The whole trouble comes from the fact that there is so much tinkering with software. It is not made in a clean fabrication process." When I found out that this remark was shocking to some of my scientific colleagues, I elaborated on the idea with the provocative saying, "What we need is *software engineering*."

This remark caused the creation of the expression "software engineering," which seemed to some to be a contradiction in terms, to be stuck in the minds of the members of the group. In the end, in late 1967, the study group recommended that we hold a Working Conference on Software Engineering, and I was made chairman. I not only had the task of organizing the meeting (which was held from October 7 to October 10, 1968, in Garmisch, Germany), but I had to set up a scientific program for a subject that was suddenly defined by my provocative remark.

---

1. Dr. Bauer originally wrote this paper as an introduction to a 1993 IEEE tutorial: *Software Engineering: A European Perspective*, R.H. Thayer, and A.D. McGettrick, eds., IEEE Computer Society Press, Los Alamitos, CA, 1993.

I enjoyed the help of my co-chairmen, L. Bolliet from France and H.J. Helms from Denmark. In addition, I had the invaluable practical support of the program committee members, A.J. Perlis and B. Randell in the section on design, P. Naur and J.N. Buxton in the section on production, and K. Samuelson, B. Galler, and D. Gries in the section on service.

Among the 50 or so participants, E.W. Dijkstra was dominant. Not only did he make cynical remarks like, "The dissemination of error-loaded software is frightening", and, "It is not clear that the people who manufacture software are to be blamed. I think manufacturers deserve better, more understanding users," but he also said, at this early date, "Whether the correctness of a piece of software can be guaranteed or not depends greatly on the structure of the thing made." He had very fittingly named his paper "Complexity Controlled by Hierarchical Ordering of Function and Variability," introducing a theme that followed his life over the next 20 years. Some of his words have become proverbs in computing, like, "Testing is a very inefficient way of convincing oneself of the correctness of a program."

With the wide distribution of reports from the Garmisch Conference and in a follow-up conference in Rome, from October 27 to October 31, 1969, it happened that not only the term "software engineering" but also the idea behind this term became fashionable. Chairs were created, institutes were established (although the one that the NATO Science Committee had proposed did not come about because of reluctance on the part of Great Britain to have it organized on the European continent), and a great number of conferences were held.

The tutorial nature of the papers in this book is intended to offer readers an easy introduction to the topics and indeed to the attempts that have been made in recent years to provide them with the *tools,* both in a handcraft and an intellectual sense, which allow them now to honestly call themselves *software engineers.*

Friedrich L. Bauer, PhD
Professor Emeritus
Technische Universität München (TUM)
Germany

In 1989, I met Dr. Friedrich L. Bauer, Professor Emeritus, Universität München, when I delivered a software engineering seminar in Munich for the IEEE. Professor Bauer later provided me with the story detailing how he came to name what we now call software engineering. I reprinted the story as an Honorary Foreword by Professor Bauer. Professor Bauer recently passed away (2015) at the age of 90. RHT

Preface

Software engineering requirements is the process of determining what is to be produced in a software system. It has the widely recognized goal of determining the needs for, and the intended external behavior of, a system design.

University students as well as candidates for the IEEE Computer Society Certificate of Proficiency exam in *software requirements* need to focus on the following eight subareas of the requirements knowledge areas (KAs) [*www.computer.org/web/education/certifications/*2015]:

1. *The knowledge and ability to establish* effective functional and non-functional requirements that are complete, concise, correct, consistent, testable, and unambiguous.

2. *The knowledge and ability to employ* appropriate requirements elicitation techniques to identify requirements.

3. *The knowledge of the skills necessary to design* a set of software models to detect and extract hidden requirements and to clearly define system functional requirements.

4. *The knowledge and ability to conduct* requirements analysis and prioritize needs.

5. *The knowledge and ability to perform* requirements engineering in the context of software development life cycles and processes.

6. *The knowledge and ability to employ* various requirements validation techniques.

7. *The knowledge and ability to critically evaluate* requirements to identify defects.

8. *The knowledge of the common techniques used to manage* the requirements effort effectively.

This book makes maximum use of SWEBOK—a very impressive document—and should be read by anyone studying software engineering.

2. Important Information for the Reader

Regarding references to SWEBOK in this book, when the second edition of the SWEBOK *Guide* is referenced, it is labeled [SWEBOK 2004]; accordingly, when the third edition of the SWEBOK *Guide* is referenced, it is labeled [SWEBOK 2014]. *A review of both of these books shows that the technical material in these two volumes is many times, word for word, identical.*

I need to point out that SWEBOK 2004 is, for all practical purposes, not copyrighted. In contrast, SWEBOK 2014 is copyrighted. Therefore, it is expedient for me to show my references, when identical in both SWEBOKs, as [SWEBOK 2004] as long as I comply with IEEE's usage limitations.

Another good reason for using SWEBOK 2014 (i.e., SWEBOK 2004) as a reference is its usefulness as an IEEE PSEM Certification exam study guide. Question writers tend to use the same wording as the question source. Since the question source is SWEBOK 2014, this would give the exam taker a "leg-up" on passing the IEEE PSEM Certification exam.

To accommodate both groups (university students as well as candidates for the IEEE Computer Society Certificate of Proficiency exam in software requirements), a software engineering principle that is not included in SWEBOK 2014 and is <u>not</u> likely to produce an exam question is marked with the following statement: "*Note: SWEBOK does not include (to be filled in) in the SWEBOK guide.* The certificate candidate is free to skip this entry. The university student should probably not.

This small book is divided into several parts.

1. The first part presents an analysis of the appropriate software engineering knowledge areas (KAs) followed by an explanation of the material related to the topics contained in the SWEBOK.

2. Part 2 contains three additional articles from technical requirement sources:

    a. The first is a software requirements tutorial by Dr. Pete Sawyer, Lancaster University, United Kingdom, intended to improve understanding of software requirements. Sawyer authored one of the reference books recommended by the IEEE Computer Society for potential exam takers.

    b. The second paper presenting the subject of software requirements "traceability" was written for an earlier software engineering tutorial presented by Dr. James D. Palmer, Professor Emeritus, George Mason University.

    c. The third paper is an excellent report published by Wikipedia explaining "prototyping."

3. Part three contains two university level requirements standards: (1) a ConOps document standard, and (2) a software requirements specification standard.

    These classroom standards should not be used to satisfy a commercial software engineering contract. Nevertheless, they can be used as (1) an educational tool and (2) a standard for students to use when preparing a classroom software requirements specification.

4. Part four includes twenty sample exam questions that should help both certification exam takers and university students.

5. Part five contains an index of key requirements terms, documents, tools, authors, and contributors.

Richard Hall Thayer, PhD, CSDP
Life Fellow of the IEEE
Member of the IEEE Computer Society Golden Core
Emeritus Professor of Software Engineering,
   Sacramento State University, California

Acknowledgments

I first want to thank my wife Mildred for her high degree of tolerance as I worked seven days a week on this manuscript. Without her help, support and dinners delivered to my work areas, this book could never have been completed.

I want to thank Ellen Sander for performing copy editing, Jon Digerness of North Coast Graphics for providing me with the comic illustrations, and Jim Tozza for giving me hardware and software support.

In addition, I want to thank Steve Tockey for providing me with numerous tips about the Computer Society exam specification process in order to maximize the usefulness of our software engineering textbook and SWE guidebook, and Melville (Mel) Piercey of Copy Plus for providing cover artwork and designing and drawing the engineering chapter graphics.

Finally, I want to thank our little dog Maxwell (a.k.a. Max, Maxcito or Speedy) who kept me company in the evening hours when everybody else had gone to bed.

A happy Max says that:

This is a Terrrrrrific book. I chewed on a copy, and it was very tasty.

A Note to Our Readers

One of the advantages of using a "print-on-demand" (POD) publishing service is the ability to make manuscript changes relatively easily when errors or improvements are identified.

The authors encourage you to identify and send potential errors or suggested improvements to the e-mail address listed below. I do not guarantee to make all the changes identified, but I do promise to review and seriously consider all recommendations.

Disclaimer

While I have more than 50 years of software engineering experience, including university teaching, I am not a technical expert in every component of software engineering. To make up for this shortcoming, I have made extensive use of material written by subject matter experts and papers (many posted on the web) as source documents.

Every effort has been made to make this software engineering reference as complete and accurate as possible. However, I can make no representation or warranties with respect to accuracy or completeness of the contents of this book and specifically disclaim any implied warrantee of merchantability or fitness for a particular purpose. The advice and strategies contained herein may not be suitable for your situation. If in doubt, you should consult with a professional software engineer. Where appropriate, neither I nor the printer will be liable for the loss of profit or other commercial damages including, but not limited to, special, incidental, consequential, or other damages [IEEE Press disclaimer].

Please keep me posted.

Richard Hall Thayer, PhD, CSDP
thayer@csus.edu

Chapter 1
Fundamentals of Software Engineering Requirements

This chapter is a textbook and reference guide introducing the concepts and common problems associated with software engineering rquirements (SER) (a.k.a. software requirements engineering or just requirements engineering). This text can be used either to supplement a university course in software engineering (SWE) or as a reference guide to aid individual software engineers in passing the IEEE Professional Software Engineering Master Certification exams.

Missing, incomplete or inaccurate software requirements are major issues in SWE. Better quality in both the software development process and the software product can be achieved if SWE methods and tools for gathering, modeling and analyzing user requirements are more effective and robust, and are codified in practice. Therefore, SER has emerged as an "engineering" approach, replacing what used to be called "requirements analysis and specification."

(This chapter uses the same general outline (paragraph headings) as defined in Chapter 1, Software Requirements, Guide to Software Engineering Body of Knowledge (SWEBOK), IEEE Computer Society, 2014). This makes comparison of the two chapter much easier.

INTRODUCTION

A SWE project contains two foundation documents. These documents must be correctly structured and written in order for the project to succeed.

(1) *Software requirements specifications (SRS)* — This chapter details the subject of SRS.

(2) *Software engineering project management (SEPM) plan* — A project management plan is discussed later in the book.

The *requirements specifications.* The software requirements specifications are the technical documents used by the designers to develop the architectural and detailed software engineering design. They describe the to-be-delivered software system. Includes reparation of the facilitates and the development of test documents and user manuals during the requirements phase.

The *project plan* is part of *SWE project management.* The plan documents the expected cost and schedule of the project that is responsible for delivering the finished software system on time and within cost, the type of engineers and other staff that are needed for the project and their schedules, and methods for monitoring and controlling project progress.

The *project manager* is responsible for the estimation and re-estimation of cost and schedule based on the software requirements. The *technical manager*, usually called the *system engineer*, is responsible for the technical quality of the requirements, controlling the changes to the requirements, and negotiating requirements changes between the customer and the developer.

The *requirements documents* explain *what* needs to be constructed and the *design* explains *how* it should be accomplished. However, it is often difficult to separate *what* from *how.* For example, to a software engineer the *system design document* is a requirement—it explains "*what*" versus "*how*"—as explained by the *software requirements specification.*

The *software requirements knowledge area* (KA) is concerned with the *elicitation, analysis, specification, validation,* and *management* of software requirements and requirements specifications. It is widely acknowledged within the software industry that SWE projects are critically vulnerable when these activities are poorly performed.

Because software requirements analysis is so similar to engineering processes, software engineering requirements (SER) has emerged as an "engineering" approach replacing what used to be called *requirements analysis and specifications.* This chapter will use the term "software engineering requirements" instead of the term "software requirements analysis and specifications." However, I will still use the independent terms "requirements analysis" and "requirements specification" when necessary and appropriate.

This KA nomenclature is broadly compatible with the SER sections contained in IEEE Std. 12207.1-96. An inherent risk of the proposed nomenclature is that a waterfall-like process may be inferred in the development of a software product. Nevertheless, a waterfall-like process is useful for providing an understanding of the various knowledge sub-areas of software requirements and their relationships to other SWE processes.

Software requirements KA engineering contains the following seven subareas [SWEBOK 2004]:

(1) *Software Requirements Fundamentals* include definitions of software requirements and the major types of requirements: product vs. process, functional vs. nonfunctional, emergent properties, etc.

(2) *Software Requirements Process* introduces the process itself, orienting the remaining five subareas and showing how requirements engineering dovetails with the other SWE processes. It describes process models, process actors, process support and management, and process quality and improvement.

(3) *Software Requirements Elicitation* is concerned with the origin of software requirements and how software engineers can identify them. It includes requirements sources and elicitation techniques.

(4) *Software Requirements Analysis* is concerned with the process of analyzing requirements to discover software boundaries and how they interact with their environment.

(5) *Software Requirements Specifications* typically refer to the production of a document that describes the software requirements that can be systematically reviewed, evaluated and approved.

(6) *Software Requirements Validation* (a subset of verification and validation, generally called V&V) is concerned with the process of examining the requirements documents to ensure that they are defining the right system (that is, the system that the user expects).

(7) *Software Requirements Management* concerns management activities and management roles belonging to requirements activities. There are two types of managers: project managers and system engineers—each has partial responsibility for creating software requirements specifications.

This chapter uses, as a manuscript outline, the paragraph headings expressed in "Software Requirements," Chapter 1, of *Guide to Software Engineering Body of Knowledge* (SWEBOK) IEEE Computer Society, 2014. This makes it easier for the reader to analyze the SWEBOK software engineering requirements processes.

SOFTWARE ENGINEERING REQUIREMENTS KNOWLEDGE AREA

The software engineering requirements (SER) knowledge area (KA) is the process of determining the requirements that need to be developed to satisfy a project agreement or contract. Figure 1 provides a top-level decomposition and breakdown of the SER KA. Figure 2 shows a sequence of software development phases, their relationships with each other, and the major software product resulting from each phase. The phases and products associated with SER are marked with a "star."

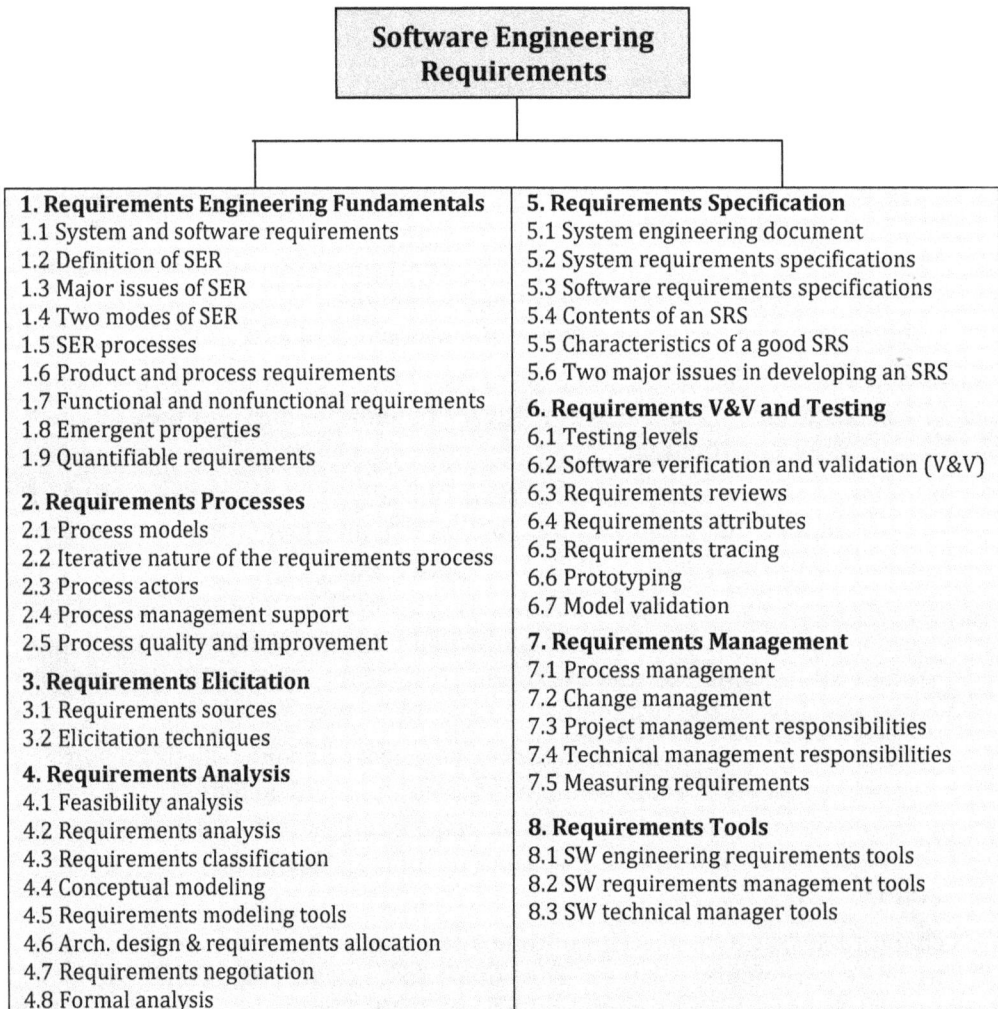

```
          ┌──────────────────────────┐
          │  Software Engineering     │
          │      Requirements         │
          └──────────────────────────┘
```

1. Requirements Engineering Fundamentals	5. Requirements Specification
1.1 System and software requirements	5.1 System engineering document
1.2 Definition of SER	5.2 System requirements specifications
1.3 Major issues of SER	5.3 Software requirements specifications
1.4 Two modes of SER	5.4 Contents of an SRS
1.5 SER processes	5.5 Characteristics of a good SRS
1.6 Product and process requirements	5.6 Two major issues in developing an SRS
1.7 Functional and nonfunctional requirements	**6. Requirements V&V and Testing**
1.8 Emergent properties	6.1 Testing levels
1.9 Quantifiable requirements	6.2 Software verification and validation (V&V)
	6.3 Requirements reviews
2. Requirements Processes	6.4 Requirements attributes
2.1 Process models	6.5 Requirements tracing
2.2 Iterative nature of the requirements process	6.6 Prototyping
2.3 Process actors	6.7 Model validation
2.4 Process management support	**7. Requirements Management**
2.5 Process quality and improvement	7.1 Process management
	7.2 Change management
3. Requirements Elicitation	7.3 Project management responsibilities
3.1 Requirements sources	7.4 Technical management responsibilities
3.2 Elicitation techniques	7.5 Measuring requirements
4. Requirements Analysis	
4.1 Feasibility analysis	**8. Requirements Tools**
4.2 Requirements analysis	8.1 SW engineering requirements tools
4.3 Requirements classification	8.2 SW requirements management tools
4.4 Conceptual modeling	8.3 SW technical manager tools
4.5 Requirements modeling tools	
4.6 Arch. design & requirements allocation	
4.7 Requirements negotiation	
4.8 Formal analysis	

Figure 1: Hierarchy of the SW engineering requirements KA

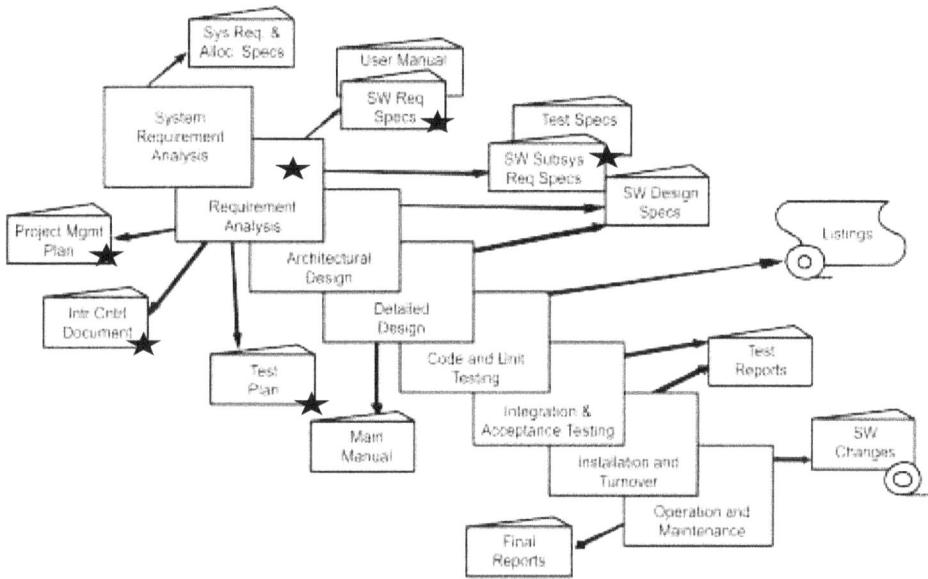

Figure 2: Sequence of software development phases and products

1. Software Engineering Requirements Fundamentals

1.1 System and software requirements.

Software requirements usually—but not always—start out as part of the system requirements specifications.

In this topic, a "system" means *"an interacting combination of elements used to accomplish a defined objective".* A system can include hardware, software, firmware, people, information, techniques, facilities, services, and other supporting elements.

System requirements are superior to software requirements in the requirements hierarchy; they are the requirements for the system as a whole. In a system containing software components, software requirements are *derived* from system requirements.

System requirements are sometimes referred to as "user requirements" in the body of requirements literature. However, "user requirements" is used in a restricted manner in this volume, referring to the requirements of the system's customers or end users. System requirements, by contrast, encompass user requirements, requirements of other stakeholders (such as regulatory authorities) and requirements without an identifiable human source [SWEBOK 2004].

1.2 Definition of Software Engineering Requirements

(1) The science and discipline concerned with analyzing and documenting software requirements is synonymous with "software requirements," and "software requirements analysis." SER involves transforming system requirements into a description of software requirements, performance parameters and a system configuration with an iterative process of definition, analysis, tradeoff studies, and prototyping.

(2) A *software capability* employed by a user to solve a problem or achieve an objective.

(3) A *software capability* that must be met or possessed by a system or system component to satisfy a contract, specification, standard, or other formally imposed document [IEEE Std. 610.12-1990].

(4) A *problem definition* that determines the real-world needs and constraints of a software system by analyzing the system requirements that have been allocated to software.

1.3 Major issues of SER. *Note:the SWEBOK does not include "major issues" in the guide.*

A widely acknowledged requirements issue — Customers expect to transition from *concepts* to *code*. They neither recognize the need nor take the time required to develop a software requirements specification or software design.

In the early 1980s, Dr. Barry Boehm identified nine more SER issues involved with requirements elicitation.

(2) *It is not always clear who the project customers or users are* — As strange as it may seem to some, and particularly to new software engineers, it may or may not always be clear who the customers are. One of the early activities of project management is to ascertain precisely who speaks for the user, who has the authority to issue contract change notices and who has the authority to make payments. Classically, these roles are covered by "the customer"; but which customer?

(3) *Customers/users cannot always state their requirements clearly or completely* — For the requirements to be effective, the customers, or their surrogate, must always state their requirements clearly, completely and correctly. Some customers find this difficult to do. Two of the current processes used today to improve this relationship between the customer and the designer are the *ConOps document* and *requirements prototyping.*

(4) *Customers/users do not know what they want; they only know what they do* — Customers frequently do not know their project goals, which makes it difficult for them to articulate a requirements specification.

(5) *What customers/users say they want may not be what they need* — There may be a conflict between what customers say they want and what they actually need.

(6) *Customers'/users' concept of a solution may not solve the software design problem* — Customers will often propose a solution that does not solve the SW problem without being specific about the initial problem. If the customer solution were followed through to the end, the project would have to be repeated to solve the problem.

(7) *Customers/users will have expectations that may be unreasonable or, more importantly, completely unknown to the engineer* —The end user may have expectations that are unreasonable to the software engineer, along with other expectations that are not communicated.

(8) *Customers do not like or are unwilling to review the finished requirements specifications* — Customers do not take the necessary time to review the SRSs after they are written to ascertain their accuracy.

(9) *Not all customers or users talk to or agree with one another, so the software engineer must talk to all of them* —It is possible for a system to have more than one customer and for the customers to not communicate with one another. It will become the engineer's responsibility to require that customers/users talk to each other to reach a common agreement about requirements specifications.

(10) *Customers/users may change* — The customers or end-users may change over time. This is particularly common in the aerospace industry, in which the customer is the Department of Defense or the Department of Energy.

A real example—after working for one year on a contract, a software developer visits the customer's plant to discuss a contractual problem and discover that the person who originally contracted for the software product is no longer employed and the current person in charge does not want the product. To have avoided this issue, thesoftware developers should have maintained constant contact with their customers.

1.4 Two modes of SER.

(1) *First*, the operational requirements are defined to describe the needs of the customer from the customer's viewpoint. The very first requirements from the customer are generally system requirements from which software requirements are derived. *Note: SWEBOK does not recognize this technical approach.*

Customer system requirements are very often operational requirements involving operational parameters (e.g., speed, performance, reliability, and capacity). A major tool for defining and documenting the operational requirements is the *concept of operations* (ConOps) document (*See also Paragraph 5.1*).

(2) *Second*, the technical requirements describe what the system should do from the technical viewpoint. The operational requirements need to be transformed intó technical requirements, which is a cooperative effort between the customer and SWE personnel.

1.5 Software engineering requirements processes.

SER can be partitioned into five different functions (*see Figure 3*). A definition for each function is provided in each section of this chapter.

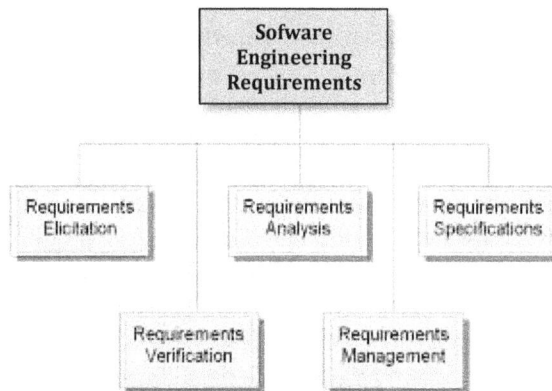

Figure 3: Process of software engineering requirements

1.6 Product and process requirements.

The description of the software product in SER is frequently portioned into *product* requirements and *process* requirements.

Product requirements apply to the product to be developed and include what the system does: data or command inputs or outputs to the system, speed and memory required by the system, quality metrics (e.g., reliability, maintainability and security), limits on the design's freedom such as interfaces (keyboard or mouse), language, and accuracy of computations.

Process requirements apply to the activities associated with enabling the creation of a product.

Some of the processes listed here are needed to produce a software system: project management, configuration management, verification and validation, and software quality. Process requirements are normally defined in a contractual statement of work (SOW) or program plan—*not in the SRS. The exception is when the customer has a "design constraint" as a requirement.*

1.7 Functional and nonfunctional requirements.

There are five general types of requirements: *functional requirements* and four *nonfunctional requirements—performance, external interface, design constraints,* and *quality attributes*. A sixth type, *other*, may be added if a system contains requirements that do not fit neatly into one of the first five types [IEEE Std. 830-1998].

(1) **Functional requirements** — A *functional requirement* is a system or software requirement that specifies a function that a system, software system or component must be capable of performing. Functional requirements define system behavior, which is the fundamental process of transformation that the system's software and hardware components must perform on inputs to produce the required outputs. Functional requirements should define the fundamental actions that must take place in the software, enabling it to perform when accepting and processing the inputs and when processing and generating the outputs.

(2) **Performance requirements** — *Performance requirements* specify a performance characteristic that a system or system component must possess—typically speed, volume or accuracy. For example, some static numerical requirements may include:

    a.   Number of customer contacts to be supported.

    b.   Number of simultaneous users to be supported.

    c.   Number of files and records to be handled.

    d.   Size of tables and files that must be handled.

    e.   Processing speed for an embedded computer system.

Dynamic numerical requirements may include the number of transactions and tasks and the amount of data to be processed within certain times for both normal and maximum workload conditions.

All requirements should be in *measurable* terms—for example, "95% of the transactions shall be processed in less than one second", rather than, "the operator shall not have to wait for the transaction to complete." All requirements must be validated (tested) as part of system development, although unmeasurable requirements may be difficult or impossible to test.

(3) **External interfaces** — A *external interface* should be a detailed description of all inputs into and outputs from the software system. It should complement the interface descriptions and should not repeat information located there.

It should include both content and format as follows: [IEEE Standard 830-1998]:

a. Name of item.

b. Description of purpose.

c. Source of input or destination of output.

d. Valid range, accuracy and/or tolerance.

e. Units of measure.

f. Timing.

g. Relationships to other inputs/outputs.

h. Screen formats/organization.

i. Window formats/organization.

j. Data formats.

k. Command formats .

l. End messages.

(4) **Design constraints** — A *design constraint* is any requirement that affects or constrains the design of a software system or software system component. Table 1 lists several examples of design constraints. *Performance requirements* and *quality attributes* may also be considered design constraints.

Table 1: Examples of system design constraints

Programming language constraint	Must use Ada.
Maximum power usage constraint	Maximum electric power that the product may use.
Human computer interface constraint	Menu system for system interfaces.
Computer resource utilization constraint	Uses no more than a specified fraction of CPU cycles, communications bandwidth, etc.
Incorporated database constraint	Must use a specified database management system.

(5) **Quality attributes** — A *quality attribute* specifies the degree of an attribute that affects the quality the software must possess. Some quality attributes are included in Table 2.

(6) *Other requirements* — Other requirements do not fit within the basic types of requirements and thus fall under the "miscellaneous requirements" category:

    a. Data definition and database requirements.

    b. Installation and acceptance requirements during operation and maintenance site functions.

    c. User documentation requirements.

    d. User operation and execution requirements.

Table 2: Examples of software quality attribute requirements

Maintainability	The average effort required to locate and fix a software failure.
Reliability	The probability that the software will perform logical operations in the specified environment without failure.
Safety	The probability that a system does not lead to a state of endangering human life or the environment
Security	The protection of computer hardware and software from accidental or malicious access, use, modification, destruction, or disclosure; the probability that the system can be made secure for a predetermined amount of time.
Survivability	The probability that the system will continue to perform or support critical functions when a portion of the system is inoperable.
User friendly	The degree of ease of use or learning of a system.

1.8 Emergent properties.

Some requirements represent *emergent properties* of software—that is, requirements that cannot be addressed by a single component but that depend on the interoperation of software components for their interoperability. For example, in a modern-day automobile control system, the system needs to interact with the automotive components to determine mileage driven, gas tank capacity, gas used, perhaps age and sobriety of driver, amount of air in tires, headlight brightness, and so forth.

Emergent properties are crucially dependent on the system architecture and are sometimes not obvious until determining how the components interact with one another.

There are two types of emergent properties [Sommerville 2007]:

(1) **_Functional emergent properties_** — These appear when all parts of a system work together to achieve an objective. For example, a bicycle has the functional property of a transportation device once it is assembled from various components.

(2) **_Nonfunctional emergent properties_** — Examples are reliability, performance, safety, and security. These relate to the behavior of the system in its operational environment. They are often critical for computer-based systems, as failure to achieve some minimally defined level in these properties may make the system unusable.

1.9 Quantifiable requirements.

Software requirements should be stated as clearly and as unambiguously as possible and, where appropriate, as quantitatively as possible. It is important to avoid vague and unverifiable requirements, which depend for their interpretation on subjective judgment ("the software shall be reliable"; "the software shall be user-friendly"; "the system shall be real time"). This is particularly important for nonfunctional requirements. Fortunately, the relatively "new" process called "concept of operations (ConOps)" can make use of these rather vague descriptions. (*See Section 5.1 for an explanation of ConOps.*) [SWEBOK 2004].

> *It should be noted that the effect of a software requirements specification is to constrain the system architecture.*

2. Requirements Processes

This section introduces the *software requirements processes*, orients the remaining five subareas, and shows how the requirements process dovetails within the overall SWE process. Figure 2 reflects a software development life cycle that includes all the major SWE products and their relationships with each other.

As stated earlier, there are *five major SER components*.

1. **_Software requirements elicitation_** — Processing through which customers (buyers and/or users) and developers (contractors) of a software system discover, review, articulate, and understand their requirements.

2. **_Software requirements analysis_** — Reasoning and analyzing customers' and users' needs to arrive at a definition of software requirements.

3. **_Software requirements specification_** — Documentation that clearly and precisely records each requirement of the software system.

4. **_Software requirements verification_** — Assuring that software requirements specifications are in compliance with system requirements, con-

form to document standards of the requirements phase and provide an adequate basis for the architectural design phase.

5. ***Software requirements management*** — Planning and controlling the requirements elicitation, analysis and verification activities.

Figure 4 is a graphic representation of a typical SWE life cycle model. In many cases, this is called the *waterfall chart* (which will be discussed later in the book).

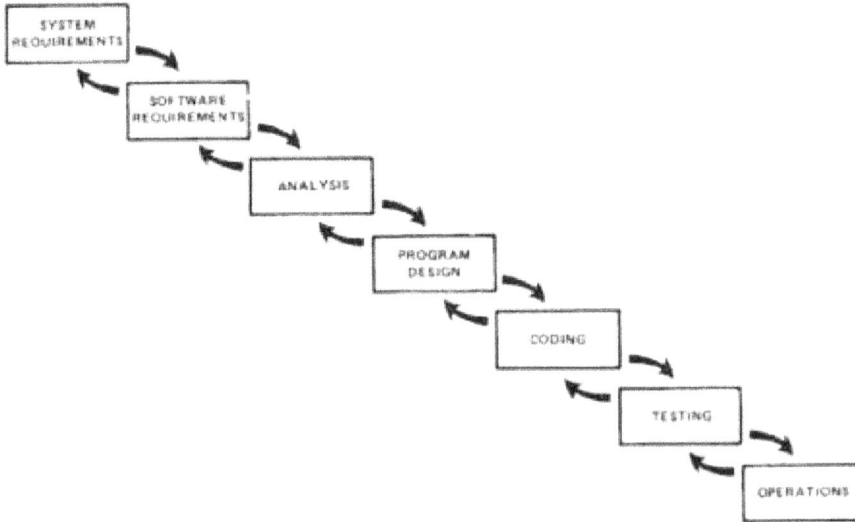

Figure 4: Waterfall chart (credited to W.W. Royce)

In this waterfall chart, project planning is executed during the requirements phase. Test plans (and some early test specifications), particularly functional tests, can also be applied at this time. Because functional requirements must be testable, developing a test plan is a useful early verification technique. Interface requirements specifications are also developed at this time.

2.1 Process models.

In the software requirements process model the SER:

(1) Is a process initiated at the beginning of a project that continues to be refined throughout the life cycle.

(2) Identifies software requirements as configuration items and manages them using the software configuration management system.

(3) Needs to be adapted to the organization and project context.

In particular, the process model is concerned with how the activities of *elicitation, analysis, specification,* and *validation* are configured for different types of projects and constraints. The process model also includes activities that provide input into the requirements process, such as marketing and feasibility studies [SWEBOK 2004].

2.2 Iterative nature of the requirements process.

In practice, it is usually impractical to implement the requirements process as a linear, deterministic process in which software requirements are elicited from the stakeholders, baselined, allocated, and handed over to the software development team. It is certainly a myth that the requirements for large software projects are ever perfectly understood or perfectly specified. Rather, the requirements process retraces even as it moves forward, correcting as it develops. *Note: SWEBOK does not cover this aspect of the waterfall chart.*

> *Dr. Winston Royce, the developer of the waterfall chart, portrayed "his" waterfall chart as both a forward looking flow of data (as each activity refines the project data) and a reverse flow of project data (to rework an earlier effort). Dr. Royce portrayed a "reverse" flow on his waterfall chart (called "iteration") to account for the return from a SWE activity to a previous activity to correct an error or to make a much-needed improvement in the overall system (see Figure 4).*

It is important to recognize the inevitability of change and to take steps to prepare for its effects. Change has to be managed by ensuring that proposed changes go through a defined review and approval process and that requirements tracing, impact analysis, and software configuration management are applied. Hence, the *requirements process* is not merely a front-end task in software development but spans the entire software life cycle. In a typical project, the software requirements activities evolve over time from elicitation to change management.

2.3 Process actors.

This topic introduces the roles of those who participate in the requirements process. There are many people called "stakeholders" involved besides the requirements specialist, each one with a "stake" in the software. This process is interdisciplinary, and the requirements specialist needs to mediate between the domains of the stakeholder and software engineers.

Although the stakeholders will vary across projects, they nearly always include users, operators and customers. A *stakeholder* is defined as any person or organization that can be positively or negatively affected by or cause an impact on the actions of a company, government or organization. Stakeholder analysis in conflict resolution, project management and business administration refers to the process of identifying the individuals or groups that are likely to affect or be affected by a proposed action, and sorting the personnel according to their effects on the action and the impact the action will have on them [http://en.wikipedia.org /wiki/Stakeholder_analysis].

Typical examples of software stakeholders involved in SER might include (but are not restricted to) [SWEBOK 2004]:

(1) **Acquisition manager** — An individual who plans, organizes and develops program management techniques and establishes internal controls to manage acquisition programs and support functions. This is normally a "customer" of the software developer.

(2) **Customers** — This group comprises those who have commissioned the software or who represent the software's target market.

(3) **Users** — This group comprises those who will operate the software. It is often a mix of people with different roles and requirements.

(4) **Project manager** — An individual (part of the development organization) who is responsible for planning, organizing, staffing, directing, and controlling the software project.

(5) **Market analysts** — A mass-market product will not have a commissioning customer, so a marketing group is often needed to establish the market needs and to act as proxy customers.

(6) **Regulators** — Many application domains such as banking and public transport are regulated. Software in these domains must comply with the requirements of the regulatory authorities.

(7) **Software developer** — An engineer or other qualified person (usually a software engineer) who develops, designs, programs, tests, and delivers a software system to satisfy a set of software requirements.

It will not be possible to satisfy the requirements perfectly for every stakeholder. It is the software system engineer's job to negotiate trade-offs that are both acceptable to the principal stakeholders and that fall within budgetary, technical, regulatory, and other constraints. Prerequisites for these trade-offs include identification of each stakeholder, analysis of the nature of their "ownership", and specification of their requirements.

2.4 Process management support.

This paragraph introduces the project management resources required and consumed by the requirements process. Its principal purpose is to make the link between the process model activities identified in Paragraph 2.1 and the issues of cost, human resources, training, and tools [SWEBOK 2004]. (*See also Paragraph 8.*)

2.5 Process quality and improvement.

This topic is concerned with the assessment of the quality and improvement of the requirements process. Its purpose is to emphasize the key role that the requirements process plays in terms of the cost, the timeliness of a software product and the customer's satisfaction with it (the three elements of SWE project success). It will help to orient the requirements process with quality standards and process improvement models for software and systems engineering [SWEBOK 2004].

> *A basic principle of SWE is that to develop a software quality product, one must have a quality process.* [2]

3. Requirements Elicitation

Requirements elicitation is the process through which the customer (buyer and/or user) and developers (contractors) of a software system discover, reveal, articulate, and understand the customer's requirements. Requirements elicitation is concerned with the origin of software requirements and how the software engineer collects them.

Elicitation is the first stage in formulating an understanding of the problem that the software is required to solve. It is fundamentally a human activity and is the point at which the stakeholders are identified and relationships are established between the development team and the customer. Elicitation is variously termed "requirements capture," "requirements discovery," and "requirements acquisition."

One of the tenets of good SWE is effective communication between the users and software engineers. Before development begins, requirements specialists may form the conduit for this communication. They must mediate between the domain of the software users (and other stakeholders) and the technical world of the software engineer [SWEBOK 2004].

Understanding these issues can mitigate them, increasing the probability that the software product will be delivered on time and within budget constraints, and will meet the user's needs wherever possible.

3.1 Requirements sources.

The system requirements specification is one of the primary sources of software requirements specifications. A team of system engineers will perform elicitation studies to determine the total number of requirements needed for a system. Together with software system engineers, they will *partition* the system requirements into detailed, testable software specifications. Those specifications that can best be completed by software requirements are then allocated to the software subsystem.

---

2. An observation made by Richard (Dick) Fairley during a meeting in 2005.

Requirements have many sources in a typical SWE project, and it is essential that all potential sources be identified and evaluated for their impacts on the project. Software engineers should be aware of the various sources of software requirements and the frameworks for managing them. The nine primary available sources of requirements information are listed below. (The topics are self defining.) *Note: SWEBOK uses a different set of requirements sources.*

(1) System requirements specifications.

(2) Statement of work (SOW) and procurement specifications.

(3) Customer prepared needs documents.

(4) ConOps documents.

(5) Observations and measurements of the current system.

(6) Interviews with customers and users.

(7) Current system documentation.

(8) Feasibility studies.

(9) Models and prototypes.

3.2 Elicitation techniques.

Once the requirements sources have been identified, the software system engineer can start eliciting requirements. Elicitation concentrates on techniques used to encourage human stakeholders to articulate their requirements. Elicitation is a very difficult area and the software engineer needs to be aware that, for example, users may have difficulty describing their tasks, may leave important information unstated or may be unwilling or unable to cooperate with design efforts.

It is particularly important to understand that elicitation is not a passive activity, and that even if cooperative and articulate stakeholders are available, the software engineer must work hard to elicit the right information. A number of good techniques exist for doing this; the five principal ones are [SWEBOK 2004]:

(1) *Interviews* — A "traditional" means of eliciting requirements. It is important to understand the advantages and limitations of conducting interviews.

(2) *Scenarios* — A valuable means for providing context to the elicitation of user requirements. Scenarios allow software engineers to provide a framework in which questions can be asked about user tasks by permitting "what if" and "how is this done" type questions. One common type of scenario is *use cases*. This is also called *conceptual modeling* because scenario notations such as use cases and diagrams are common when modeling software.

(3) **Prototypes** — A valuable tool for clarifying ambiguous requirements. A prototype is a working software system. It can act in a similar way to scenarios by providing users with a context within which they can better understand what needs to be provided. There is a wide range of prototyping techniques, from paper mock-ups of screen designs to beta-test versions of software products, with a strong overlap of their uses for requirements elicitation and for requirements validation. (*See also Chapter 4.*)

(4) **Joint application development (JAD) meeting** — A JAD meeting offers a *team-oriented approach* to the development of information management solutions that emphasizes a consensus-based, problem-solving model. By incorporating facilitated workshops and emphasizing a spirit of partnership, JAD enables system requirements to be documented more quickly and accurately than when a traditional approach is used. JAD combines technology and business needs in a process that is consistent, repeatable and effective.

In a "perfect" JAD meeting, the *executive sponsor,* who has the *ultimate authority* to make decisions about the project, should also attend. The sponsor may be the customer's project leader, the CIO, or in some cases, senior management. A facilitator works with the sponsor to start the project; it is important, however, that the sponsor makes decisions, not the facilitator.

When it works well, this technique may result in a richer and more consistent set of requirements than might otherwise be achievable. However, JAD meetings need to be handled carefully (hence, the need for a facilitator) to prevent the critical abilities of the team from being eroded by group loyalty, or the requirements from being altered to reflect the concerns of a few outspoken (and perhaps senior) group members. Neither certain abilities nor members should be favored to the detriment of others.

(5) **Observation** — Engineers learn about user tasks by immersing themselves in the environment and observing how users interact with their current software and with each other. Observation techniques are relatively expensive, but are instructive because they illustrate the many user tasks and business processes that are too subtle and complex to be easily described by their actors.

Other sources of information essential to a project are the *statement of work* and *procurement specifications* delivered by the contract. As previously noted, a *customer needs document* (the ConOps document), *measurements* of the original system, *interviews* with customers and users (a classic way of determining software requirements), and *feasibility studies* may also be available.

4. Requirements Analysis

The process of analyzing requirements is designed to accomplish the following [SWEBOK 2004]:

(1) Elaborate on system requirements to derive software requirements.

(2) Detect and resolve conflicts between requirements.

(3) Discover the bounds of the software with its environment.

(4) Perform due diligence on the potential system.

4.1 Feasibility analysis. *Note: SWEBOK does not include feasibility analysis in the exam guide.*

A feasibility study aims to objectively and rationally uncover the strengths and weaknesses of an existing business or proposed venture, opportunities and threats present in the environment, the resources required to create the software design, and ultimately the prospects for success. In the simplest terms, the two criteria to judge feasibility are required cost and value to be attained. [Justis & Kreigsmann 1979], [https://en.wikipedia.org/wiki/Feasibility_study].

A well-designed feasibility study should provide a historical background of the business or project, a description of the product or service, accounting statements, details of the operations and management, marketing research and policies, financial data, legal requirements, and tax obligations. Generally, feasibility studies precede technical development and project implementation.

A feasibility study evaluates the project's potential for success; therefore, perceived objectivity is an important factor in the credibility of the study for potential investors and lending institutions. It must therefore be conducted with an objective, unbiased approach to provide information upon which decisions can be based.

4.2 Requirements analysis.

The traditional view of requirements analysis has been that it be reduced to *conceptual modeling* using one of a number of analysis methods such as structured analysis (a.k.a. the Yourdon technique). While conceptual modeling is important, the classification of requirements is included to help inform trade-offs between requirements (requirements classification) and the process of establishing those trade-offs (requirements negotiation).

Care must be taken to describe requirements precisely enough to enable the *requirements* to be *validated*, their *implementation* to be *verified* and their *costs* to be *estimated.*

4.3 Requirements classification.

Requirements can be classified in a number of dimensions. Examples include:

(1) ***Operational requirements vs. technical requirements*** —Two types of software requirements can be defined. The first is an *operational requirement,* which describes the needs of the customer from the customer's viewpoint. The very first requirements obtained from the customer are usually *system requirements* or a *ConOps.*

A technical requirement pertains to the technical aspects that your sys

tem must fulfill, such as performance-related issues, reliability issues, and availability issues.

(2) ***Product vs. process requirements*** —*Product requirements* apply to the product or services to be developed. These include what the system does: data or command inputs or outputs to the system, speed and memory required by the system, quality metrics (e.g., reliability, maintainability and security), and limits on the design's freedom such as interfaces (keyboard or mouse), language and accuracy of computations.

Process requirements apply to the activities associated with enabling the creation of a product or service. Some examples of tasks to be performed in process requirements classification are classifying processes that are needed to produce a software system (e.g., project management, configuration management, testing, verification and validation, software quality assurance, and so forth).

(3) ***Functional vs. nonfunctional requirements*** —A *functional requirement* is a system or software requirement that specifies a function that a system, software system or component must be capable of performing. *Nonfunctional requirements* are requirements that do not directly affect the functionality of the system but are an integral part of the software specifications.

(4) ***Primary requirements vs. derived requirements*** — *Primary requirements* are imposed directly on the software by a stakeholder or some other source, and *derived requirements* are derived from one or more high-level requirements.

It is always important to know the source location of software requirements when changes need to be made.

(5) ***Priority levels*** — Software and system requirements are compartmentalized into *mandatory, desirable* and *optional* requirements, each one being characterized by such English words as "shall," "will" or "may."

"Notes" can also be used to provide auxiliary information to the software development organization. These are <u>not</u> classically considered to be "contractually binding" requirements.

Some types of priorities are often tied to requirements as guidance to software developers. Obviously, high priority requirements are "essential" while low priority requirements might be "optional." The purpose of prioritizing requirements is to identify in advance what might be dropped if the project were to fall behind in cost and schedule but the "powers-to-be" insist on adhering to the agreed cost and/or schedule.

Decisions of this type should preferably be agreed upon at the beginning of the project rather than after the project experiences financial difficulty.

(6) **Volatility vs. stability** — Some requirements will change during the life cycle of the software and even during the development process itself. Preparing an estimate of the likelihood that a requirement will change can be useful. For example, a business application will probably be more stable that a military application. Flagging potentially volatile requirements can help the software engineer to establish a design that is more tolerant of change. Other classifications may be appropriate depending upon the organization's normal practice and the SW application itself.

4.4 Conceptual modeling.

The development of *conceptual models* of real-world problems is a key to software requirements analysis. Their purpose is to aid in understanding the identified problem, rather than to initiate design of the solution. Hence, conceptual models comprise models of entities from the problem domain configured to reflect their real-world relationships and dependencies. (*Note: conceptual models can also be used in software design*)

The purpose of the conceptual model is to develop a solution to the identified problem.

Many kinds of conceptual models have been developed. The factors that influence the choice of model to be used to identify the problem include the following [SWEBOK 2004]:

(1) **Nature of the problem** — Some types of software demand that certain aspects be analyzed particularly rigorously. For example, control flow and state models are likely to be more important for real-time software than for management information software, which would usually be the more appropriate choice for data models.

(2) **Expertise of the software engineering team** — It is often more productive to adopt a modeling notation or method with which the software engineer has experience.

(3) **Process requirements of the customer** — Customers may impose their favored notation or method, or prohibit any with which they are unfamiliar. This can conflict with the previous factors.

(4) *Availability of methods and tools* — Notations or methods that are poorly supported by training and tools may *not* achieve widespread acceptance even if they are suited to particular types of problems.

4.5 Requirements modeling tools.

Below is a list of a number of popular SWE conceptual models that have existed since the beinning of software engineering. I used many of these tool taught the use of the tool in a university course, or attended a course or a lecture on their use. The described method may predate or postdate the provided date by many years. This partial list of SWE requirements models is provided as a reminder that almost all of these techniques were viewed as being "wonder" tools or methods that would be "The Tool" to use for time immemorial.

(1) *Structured analysis* was extremely popular in the 1970s and is still in limited use in the 2000s. Structured analysis consists of modeling the system using data-flow diagrams (a.k.a. a bubble chart).

Structured analysis is part of the *"structured method"*, including context diagrams, dataflow diagrams, mini-specs, data dictionaries, structured design, and structured programming, etc. Many of these techniques were developed by Edward Yourdon [1989] and are credited to him.

Strctured anslysis consisted ot the following"

a. A *[system] context diagram* is a SWE and systems engineering diagram that defines the boundary between the system, or part of a system, and its environment, showing the entities that interact with it. This diagram is a high-level view of a system. It is similar to a block diagram, as as presented in [http://en.wikipedias.org/wikiSystem context_diagram].

b. A *data flow diagram* (*DFD*) is a graphical representation of the "flow" of data through an information system. A bubble is used to represent a requirements process. The process is represented on paper as a circle (i.e., a bubble). Each bubble can be partitioned into a set of bubbles, each one representing a subset of the original process. The flow of data can be either from a bubble to a data store and then to another bubble, or can flow between bubbles. A bubble chart (a system of bubbles) can become extremely large.

c. *Data dictionaries* are needed to describe the data flows and process specifications which are needed to capture the transaction and/or transform information. For more information, see [http://en.wiki pedia.org/wiki/Structured_analysis].

d. *Mini-specs* are structured analysis tools that are used to describe the process contained within a bubble on a bubble chart. Yourdon defines a mini-spec as being no longer than two pages.

(2) An early software engineering tool development was the *Information System Design and Optimization System* (ISDOS) project, started in 1968 at the University of Michigan. This software development toolsinitiated a great deal of interest in the concept of using computer systems to help analysts with the very difficult process of analyzing requirements and developing systems. Several papers authored by Daniel Teichroew motivated an entire generation of software developers to explore the potential of automated systems development. Teichroew's Problem Statement Language/Problem Statement Analyzer (PSL/PSA) tool was a CASE tool, although it predated the term.

(3) *HIPO model* (short for *hierarchical input process output* model) is a systems analysis design aid and documentation technique developed by IBM in 1974.

The overall design of the system is documented using HIPO charts or structure charts.. Structure charts can be used to display several types of information, but are used most commonly to diagram either data structures or code structures. I used HIPO charts at Vandenberg in the 1970.

(4) The *Structured Analysis and Design Technique* (SADT), developed by Doug Ross in the mid-70s uses activity diagrams rather than dataflow diagrams. SADT also distinguishes control data from processing data.

(5) *Structured Analysis and System Specification* (SASS), developed by Ed Yourdon and Tom DeMarco in the mid-70's, is "classic" structured analysis.

(3) *Real-time structured analysis* is a variation of the structured analysis system of Yourdon and DeMarco. This SWE technology is credited to P.T. Ward and S.J. Mellor in 1985 and D.J. Hatley and I.A. Pirbhai in 1988. The conventional flow diagram is supplemented by "control flow."

(4) A *CASE (computer aided software engineering) tool* provides computer-based support for the software development process. A CASE tool is a computer-based product aimed at supporting one or more SWE activities within a software development process. A CASE environment is a collection of CASE tools and other components together with an integration approach that supports most or all of the interactions that occur among the environment components, and between the users of the environment and the environment itself.

(5) *Software through Pictures (StP)* is a CASE tool/environment developed and marketed by *Interactive Development Environments*. This tool was developed by Dr. Tony Wasserman in the late 1980s. This CASE tool used

graphic illustrations (pictures) to identify and display the software requirements.

In the early 1990s, Tony Wasserman gave my university five copies of his "Software through Pictures" system to use in our SWE graduate courses. Thank you again, Tony.

(6) An *event trace* or *tracing* is a specialized use of logging to record information about a program's execution. This information is typically used by programmers for debugging purposes and by experienced technical support personnel as a software monitoring tool to diagnose common problems with software [http://en.wikipedia .org/wiki/Tracing_(software)].

(9) The *object model* has two related but distinct meanings:

a. The general properties of objects in a specific computer programming language, technology, notation, or methodology that use objects. Such object models are usually defined using concepts such as class, generic function, message, inheritance, polymorphism, and encapsulation. There is extensive literature supporting formalized object models as a subset of the formal semantics of programming languages [http://en.Wikipedia.org/wiki/Object_model].

b. A collection of objects or classes through which a program can examine and manipulate specific parts of its world; in other words, the object-oriented interface to a service or system.

(10) The term *data model* is used in two related senses. In one sense it is a description of the objects represented by a computer, typically "real-world" objects, such as products, suppliers, customers, and orders. It is also a collection of concepts and rules used in defining data models: for example, the relational model uses relations and tuples, while the network model uses records, sets and fields [http://en.wikipedia.org/wiki /Data _model].

(11) The *Unified Modeling Language* (UML), is an object diagram which focuses on some particular set of objects and attributes, and the links between these instances. UML was originally motivated by the desire to standardize the disparate notational systems and approaches to software design developed by Grady Booch, Ivar Jacobson and James Rumbaugh at Rational Software in 1994–1995, with further development led by them through 1996,

(12) *Agile* software development was initiated in 2001. This software engineering tool describes a set of principles for software development under which requirements and solutions evolve through the collaborative effort of self-organizing cross-functional teams. (A cross-functional team is a group of people with different functional expertise working toward a

common goal. It may include people from finance, marketing, operations, and human resources departments. Typically, it includes employees from all levels of an organization. Members may also come from outside an organizatiosuppliers, key customers or consultants.)

Agile advocates adaptive planning, evolutionary development, early delivery, and continuous improvement and it encourages rapid and flexible response to change. These principles support the definition and continuing evolution of many software development methods [https://en.wikipedia.org/wiki/Agile_software_development].

(13) *Protyping* is a simple but effective way of moding a set of requirements. A prototype is an early sample, model, or release of a software product built to test a concept or process or to act as a thing to be replicated or learned

A prototype is generally used to evaluate a new design to enhance precision by system analysts and users. Prototyping serves to provide specifications for a real, working system rather than a theoretical one (See Chapter 4).

(14) *Formal modeling* uses notations based on discrete mathematics that are traceable to *logical reasoning*. These notations may be a concept model or the customer's requirements, or they may offer a compelling advantage to understanding the analysis of certain critical functions or components. (See also Paragraph 4.8.)

We do not seek to "teach" a particular modeling style or notation but rather to provide guidance with regard to the purpose and intent of modeling [SWEBOK 2004].

Note: The only reason for "boring" you with the very long (but still incomplete) list of SWE tools is to remind you that SWE tools and methods have a limited use life. Many of the tools included on this list are long gone, having been replaced with new techniques that promised to be more accurate and easier to use. These new techniques were in turn replaced with still other processes, which in turn are also being replaced, and so forth. Who is predicting when Agile will be replaced with a new technique that promises to be again more accurate and easier to use?

In all fairness, some of these processes have been more enduring than others—for example, *structured analysis, prototyping, UML,* and, who knows, maybe even *Agile.*

Software engineers (university students or working professionals) need to be able to recognize and use a *SWE modeling tool* that is either made available to them or required by a higher authority (like a university instructor). The message here is to use what is available and/or required, but remember: the usefulness of a particular tool or method is measured in years and not decades.

To use a model, it is frequently beneficial to start by building a model of the software context (a context diagram—a top-level bubble). The software context provides a connection between the intended software and its external environment. This is crucial to understanding the software's context in its operational environment and identifying its interfaces with the environment.

The issue of modeling is tightly coupled with that of methods. For practical purposes, a method is a notation (or set of notations) supported by a process that guides the application of the notations. There is little empirical evidence to support claims for the superiority of one notation over another. However, the widespread acceptance of a particular method or notation can lead to beneficial industry-wide pooling of skills and knowledge. This is currently the situation surrounding use of the Unified Modeling Language (UML) [SWEBOK 2004].

4.6 Architectural design and requirements allocation.

At some point, the architecture of the solution must be derived. *Architectural design* is derived when the requirements process overlaps with software or sys

tems design. It illustrates the impossibity of cleanly decoupling the two tasks [SWEBOK 2004]. In many cases, the software engineer acts as a software designer because the process of analyzing the requirements and elaborating on their effectiveness demands that the components be responsible for satisfying the requirements be identified. This is *requirements allocation*, i.e., the assignment of software design components, or the responsibility for satisfying a given software requirement [SWEBOK 2004].

Allocation is important to permit detailed analysis of requirements. Hence, for example, once a set of requirements has been allocated to a software component, the individual requirements can be further analyzed to discover additional requirements to determine how the component needs to interact with other components to satisfy the allocated requirements.

In large projects, allocation stimulates a new round of analysis for each subsystem. As an example, requirements for a particular braking performance for a car (braking distance, safety in poor driving conditions, smoothness of application, pedal pressure required, and so on) may be allocated to the braking hardware (mechanical and hydraulic assemblies) and an anti-lock braking system (ABS). Only when a requirement for an ABS has been identified and the requirements have been allocated to it can the capabilities of the ABS, the braking hardware and emergent properties (such as the car weight), be used to identify the detailed ABS software requirements.

Architectural design is closely identified with *conceptual modeling*. The mapping from real-world domain entities to software components is not always obvious, so architectural design is identified as a separate topic. The requirements of notations and methods are broadly the same for both conceptual modeling and architectural design [SWEBOK 2004].

4.7 Contract resolution.

Another term commonly used for this subtopic is *requirements negotiation*. This concerns resolving problems caused by conflicts that occur between two stakeholders requiring mutually incompatible features, between requirements and resources, between functional and nonfunctional requirements, or between two customers wanting to use the same completed system [SWEBOK 2004].

> In the early 1960s, the U.S. Air Force needed to build an airborne Strategic Air Command (SAC) command post. To make the effort reasonable, the project was to share both features and finances between two organizations. However, the two agencies could not agree on a number of important technical features belonging to the system. The two requirements were dynamically different—and only one agency could "win." (Can you guess which design was built?). P.S. The other agency was the NSA.

4.8 Formal analysis.

Formal methods are mathematically based approaches to software development in which a formal (mathematical) model of the software is defined. One may then formally analyze this model and use it as a basis for a formal system specification. In principle, it is possible to start with a formal model for the software and prove that a program is consistent with that model, thus eliminating software failures resulting from programming errors.

The starting point for all formal development processes is a formal system model which serves as a system specification. To create this model, one translates the system's user requirements, which are expressed in natural language, diagrams and tables, into a mathematical language that has formally defined semantics. The formal specification is an unambiguous description of what the system should do. Using manual or tool-supported methods, one can check that a program's behavior is consistent with the specification [Sommerville 2010].

5. Requirements Specification

For most engineering professions, the term "specification" refers to the assignment of numerical values or limits to a product's design goals.

Typically, software has a large number of requirements and emphasis is shared between performing the numerical quantification and managing the complexity of interactions among the large number of requirements [SWEBOK 2004].

Therefore, an SRS typically refers to the document, or its electronic equivalent, that can be systematically reviewed, evaluated and approved. For complex systems, particularly those involving substantial non-software components, as many as three different types of documents are produced.

Three of these documents are described here with the understanding that they may be combined as appropriate. For simple software products, only one of these documents is required. The formats for a system requirements specification and a software requirements specification share many common components.

5.1 System engineering document.

The current approach for a system definition document is the *concept of operations (ConOps)* document. The *ConOps* document is a bridging document between the customer's wants, needs and expectations and the technical requirements document called a "software requirements specification." The ConOps may be thought of as the "missing link" of SER [Fairley & Thayer 1997].

The sysem engineering document (the ConOps document) contains:

- A description of the current system or situation.

- A description of the needs that motivate development of a new system or modification of an existing system.

- Modes of operation for the proposed system.

- User classes and user characteristics.

- Operational features of the proposed system.

- Priorities among proposed operational features.

- Operational scenarios for each operational mode and class of user.

- Limitations of the proposed approach.

- Impact analysis for the proposed system.

In contrast to a requirements specification, a ConOps document should be written in narrative prose, using the language and terminology of the users' application domain. It should be organized so as to tell a story, and should make use of visual forms whenever possible.

Prior to the advent of the ConOps document, the software engineering specifications were usually written using the users' terms and terminology, or less frequently, using the software designer's technical terms and terminology; it was very difficult to satisfy both entities. This traditional method of specifying software requirements introduces several problems:

(1) The *acquirer* is the customer's "project manager", and usually the individual responsible for interfacing with the developer, who may not ade-

quately convey the needs of the user community to the developer, perhaps because the acquirer may not understand the users' needs.

(2) The *developer* may not be an expert in the field for which the application is being created, which inhibits communication.

(3) The *users* and *acquirer* often find it difficult to understand the requirements produced by the developer.

(4) The *developer's* requirements specification typically specifies system attributes such as functions, performance factors, design constraints, system interfaces, and quality attributes, but typically contains little or no information concerning operational characteristics of the specified system [RADC-TR-85-37, RADC-TR-175].

These issues leave the users and the acquirer uncertain as to whether the requirements specifications describe a system that will provide the needed operation capability.

A ConOps document contains a description of the current system or situation, the nature of proposed changes and/or new features of the operational concepts for the new or modified system and their justification, operational scenarios for the new system, a summary of organizational and operational impacts, and an analysis of the proposed system. The ConOps document is typically written by the user and/or customer or by some organization representing them.

In addition, the ConOps provides a place for users to state their *desires*, *visions*, and *expectations* without requiring them to provide quantified, testable specifications.

For example, the users could express their need for a "highly reliable" system and their reasons for that need, without having to produce a testable reliability requirement. It will then be left to the discretion of the technical personnel, with support from the users, to convert these vague but necessary terms into measurable quantities to create the technical requirements specifications. The ConOps is sometimes called a *system definition document.*

5.2 System requirements specifications.

Developers of systems with substantial software and nonsoftware components, like a modern airliner, often separate the description of system requirements from the description of software requirements. In this view, system requirements are specified, the software requirements are derived from the system requirements, and lastly, the requirements for the software components are specified [SWEBOK 2004].

5.3 Software requirements specifications (SRS).

Software requirements specifications establish the basis for agreement between customers and contractors or suppliers (in market-driven projects, these roles may be played by the marketing and development divisions) regarding what the software product is to do, as well as what it is not expected to do. An SRS permits a rigorous assessment of requirements before design can begin and reduces later redesign (called *rework*). It also provides a realistic basis for estimating product costs, risks and schedules [IEEE Std. 830-1998].

Organizations can also use an SRS document to develop their own project management plans, users' manuals and verification and validation and (V&V) plans. Since the requirements specification describes the delivered system, the users' manual (or at least a draft of the users' manual) can be developed and published in parallel with the requirements specification [IEEE Std. 830-1998].

The great benefit of developing the users' manual early, along with the requirements, is the effect it has on the eventual user. In many cases, the user of the software system reviews but apparently cannot understand the SRS, and consequently, will give it only a cursory glance, providing an "underinformed approval." This approval will overlook many potential SWE product disasters. However, in other scenarios, the user will usually read the users' manual with great enthusiasm and understanding, thereby approving or rejecting the manual for sound technical reasons. Since the users' manual is based on the requirements' functionality, the results are an educated acceptance or rejection of the requirements specifications [IEEE Std. 830-1998].

SRS provides an informed basis for transferring a software product to new users or new machines. It can also provide a basis for software enhancement [IEEE Std. 830-1998].

Software requirements are often written in natural language. SRSs may be supplemented by formal or semiformal descriptions. Selection of appropriate notations permits particular requirements and aspects of the software architecture to be described more precisely and concisely than by using natural language [IEEE Std. 830-1998].

The general rule is that notations should be used that allow the requirements to be described as precisely as possible. This is particularly crucial for safety-critical and certain other types of dependable software. However, the choice of notation is often constrained by the training, skills and preferences of the document's authors and readers [IEEE Std. 830-1998].

The SRS is intended for use by software users/customers, system/software engineers, and system analysts or students that need to document the results of a system/software requirements analysis. The users of this specification are encouraged to incorporate additional items into the specification to satisfy unique situations, to expand specific sections to fully describe the scope and magnitude of particular software requirements specifications, or to make other modifications to suit the design circumstances and environment [IEEE Std. 830-1998].

5.4 Contents of an SRS. (*Note: SWEBOK does not include this function in the SWEBOK guide.*)

The contents of an SRS document the conditions and capabilities needed by a user to solve a problem or achieve an objective. The set of all requirements forms the basis for subsequent development of the software system or software system components. A complete specification will include functional requirements, performance requirements, external interface requirements, design constraints, and software quality attributes. Each of these is discussed below.

(1) *Functional requirements* — Requirements specifying a function that a system or system component must be capable of performing.

(2) *Performance requirements* — Requirements that specify a perfor mance characteristic that a system or system component must possess; for example: speed, accuracy and frequency.

(3) *External interface requirements* — Requirements that specify a hardware, software or database element with which a system or system component must interface, or that sets forth constraints on formats, timing or other factors caused by such an interface. Other examples of external interfaces are: people, special purpose hardware, other systems, software from other projects, etc.

(4) *Design constraints* — Requirements that impact or constrain the design of a software system or software system component; for example: functional requirements, physical requirements, performance requirements, software development standards, and software quality assurance standards. Other examples of design constraints are: required implementation language, database integrity, limits on resources such as memory, etc.

(5) *Software quality attributes* — Requirements of software which affect its perceived value; for example, correctness, reliability, maintainability, portability, and security. Note that not all specifications are required by a contractual document. Many requirements have been identified through analysis or design as necessary to satisfy a higher-level requirement. These are often referred to as *derived requirements*.

5.5 Characteristics of a good SRS.

Every SRS document should contain the following nine quality attributes. It is recognized that in some cases these attributes cannot be completely realized, and therefore, they serve as a quality goal for the software requirements specifications:

(1) *Complete* — No requirements overlooked.

(2) *Consistent* — No set of individual requirements conflicts with any other set.

(3) **Correct** — No error exists that will affect design.

(4) **Modifiable** — Any necessary changes can be made completely and consistently.

(5) **Process-free** — Design and management requirements are excluded.

(6) **Ranked** — Placed in order of importance and/or stability.

(7) **Traceable** — There is an audit trail from requirements to tested code.

(8) **Verifiable** — A finite process exists to verify that the product meets the requirements.

(9) **Unambiguous** — There is only one semantic interpretation of the requirements.

5.5.1 Complete. An SRS is *complete* if it possesses the following qualities:

(1) Inclusion of all significant requirements, whether relating to functionality, performance, design constraints, attributes, or external interfaces.

(2) Definition of the responses of the software to all realizable classes of input data in all realizable classes of situations. Note that it is important to specify the responses to invalid as well as valid input values.

(3) Conformity to any SRS standard that applies to the SRS. If a particular section of the guide is not applicable, the SRS should include the section number and an explanation outlining why it is not applicable.

(4) Full labeling and referencing of all figures, tables and diagrams in the SRS and definitions of all terms and units of measure.

(5) Sometimes it is necessary to delay the completion of a specification until additional information can be obtained. Under these circumstances the annotation is entered "to-be-determined", i.e., TBD. When a TBD is used it needs to be accompanied by: (a) a description of the conditions causing the TBD (for example, why an answer is not known) so the situation can be resolved, and (b) a description of what must be done to eliminate the annotation.

5.5.2 Consistent. An SRS is *consistent* if, and only if, no set of individual requirements described are in conflict. There are two types of likely conflicts in an SRS:

(1) Two or more requirements might describe the same real world object but use different terms for that object. For example, a program's request for a user input might be called a "prompt" in one requirement and a "cue" in another. In another example, the user of the system might be called a "user" in one paragraph and a "customer" in another paragraph.

(2) The specified characteristics of real world objects might conflict. For example, the format of an output report might be described in one requirement as "tabular" but in another as" textual."

5.5.3 Correct. An SRS is *correct* if, and only if, every requirement stated therein is one that the software shall meet. Each requirement does <u>not</u> have errors that will create errors in the system design.

5.5.4 Modifiable. An SRS is *modifiable* if its structure and style are such that any necessary changes to the requirements can be made easily and completely.

An SRS should not contain redundant specifications. Modifiability generally requires that the same requirement does not appear in more than one place in the requirements specification. Redundancy itself is not an error but can easily lead to errors.

Redundancy can occasionally help to make an SRS more readable; however, a problem can arise when the redundant document is updated. Assume, for instance, that a certain requirement is stated in two places. At some later time, it is determined that the requirement should be altered, but the change is made in only one of the two documents. The SRS then becomes inconsistent.

Whenever redundancy is needed, the SRS should include cross-references between the redundant entries.

5.5.5 Process-free. An SRS is *process free* when it only specifies the results that must be achieved by the software, not the means of planning for or obtaining those results.

5.5.6 Ranked. A *ranking* is a relationship between a set of items such that, for any two items, the first is either 'ranked higher than', 'ranked lower than' or 'ranked equal to' the second. By reducing detailed measures to a sequence of ordinal numbers, rankings make it possible to evaluate complex information according to certain criteria.

Thus, for example, a set of software requirements can be ranked based on their importance and/or their stability. This will enable management to make informed decisions if software requirements need to be reduced in size or complexity.

5.5.7 Traceable. An SRS is *traceable* if the origin of each of its requirements is clear and if the SRS facilitates the referencing of each requirement in future development or enhancement documentation. Two types of traceability are recommended:

(1) *Backward traceability* (that is, to previous stages of development) depends upon each requirement explicitly referencing its source in the previous documents.

(2) *Forward traceability* (that is, to all documents spawned by the SRS) depends upon each requirement in the SRS having a unique name or reference number.

When a requirement in the SRS represents an allocation or a derivative of another requirement, both forward and backward traceability should be provided.

5.5.8 Verifiable. An SRS is *verifiable* if, and only if, every requirement stated therein is verifiable. A requirement is verifiable if, and only if, a finite cost-effective process exists with which a person or machine can check that the software product meets the requirements.

Examples of unverifiable requirements include statements such as:

- "The product should be user friendly", or "The product should have a good human interface". These requirements cannot be verified because it is impossible to define the terms "good" or "friendly."

- "The program shall never enter an infinite loop." This requirement is non-verifiable because the testing is theoretically impossible.

- "The output of the program shall be given within 10 seconds." This requirement is non-verifiable because "usually" cannot be measured.

An example of a verifiable statement is: "The output of the program shall be given within 20 seconds of event X, 60% of the time, and shall be given within 30 seconds of event X, 99% of the time." This statement can be verified because it uses concrete terms and measurable quantities.

If a requirement is not expressible in verifiable terms at the time the SRS is prepared, then a point in the development cycle (review, test plan issue, etc.) should be identified, at which point the requirement must be put into a verifiable form.

5.5.9 Unambiguous. An SRS is *unambiguous* if, and only if, every requirement stated therein has only one interpretation. At a minimum, this requires that each characteristic of the final product be described using a single unique term.

In cases where a term used in a particular context could have multiple meanings, the term must be included in a glossary specifically defining each meaning.

One way to avoid the ambiguity inherent in natural language is to write the SRS using a formal requirements specification language. These languages include SERM, PSL or structured English.

One major *advantage* in the use of such languages is the reduction of ambiguity. This occurs, in part, because the formal language processors automatically detect many lexical, syntactic and semantic errors. One major *disadvantage* in the use of such languages is the length of time required to learn how to properly use them.

5.6 Two major issues in developing an SRS.

5.6.1 Embedding design in the SRS. Embedding *design specifications* in the SRS unduly constrains the software design and artificially places potentially dangerous requirements in the SRS.

(1) The SRS must specify the functions that are to be performed on a particular data set during a specific time period to produce certain results at a particular location and for a specific purpose. The SRS should not normally specify design items such as:

    a. Partitioning the software into modules.

    b. Allocating functions to the modules.

    c. Describing the control or information flow between modules.

    d. Choosing data structures.

(2) It is not always practical to consider the design as being completely isolated from the SRS. Security or safety considerations may impose requirements that reflect directly onto design constraints; for example, the need to keep certain functions in separate modules. (This is called *design constraints*—as discussed in Paragraph 1.7(3).

5.6.2 Embedding project management plans in the SRS. The SRS should address the software product, not the process of producing the software product. The SRS describes the product; the software project management plan describes the project which will produce the product.

(1) *Project plans* represent an understanding between a customer and a developer about the contractual matters pertaining to the production of software (and thus should not be included in the SRS). These normally include such items as:

    a. Cost.

    b. Delivery schedules.

    c. Reporting procedures.

    d. Software development methods.

    e. Quality assurance.

    f. Validation and verification criteria.

    g. Acceptance procedures.

(2) *Project management activities* are specified in other documents, typically in a computer program development plan, software project management plan or a statement of work (SOW). The requirements for only the software product itself are given in the SRS.

An SRS should not contain requests for services. The SRS describes the results of a service and not the service itself.

6. Requirements Verification, Validation, and Testing

Figure 5, which dates back to the 1970s, is one of the oldest figures belonging to the SWE discipline. It reflects the cost of fixing a software requirement error at various phases in the software development life cycle.For example, if an error is made during the requirements phase, it might require one dollar or one hour to fix if found during the architectural design phase. It might cost twice as much if found in the detailed design phase, five times more if found in the integration and test phase, and 100 times more if found by the user in the operational phase.

The properities of verification and validation (V&V) lend themselves to the early detection of software errors.

6.1 Testing levels.

(1) *Unit testing* tests for coding errors.

(2) *Integration testing* checks for design and interface errors.

(3) *System testing* finds requirements errors as seen by the developers and frequently uses artificial test cases in order to test for unique errors and stress testing.

(4) *Acceptance testing* confirms that requirements errors surface as determined by the system user.

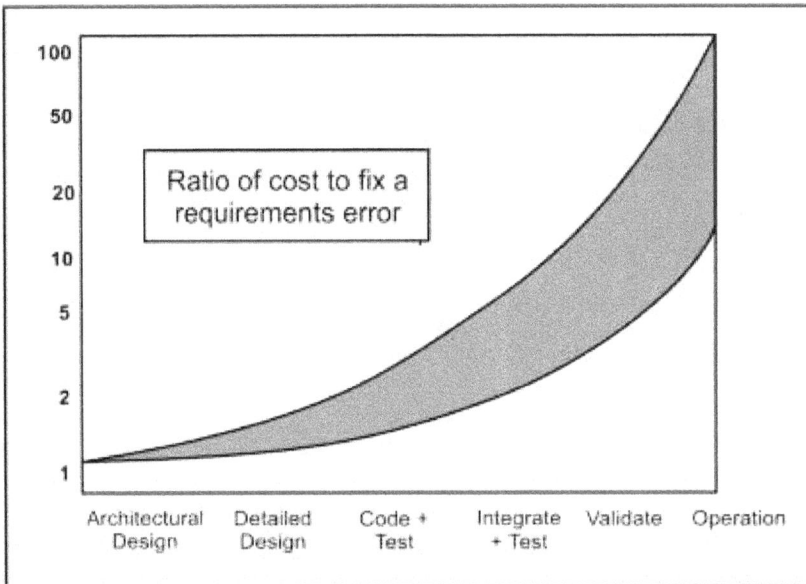

Figure 5: Errors made first are discovered last

Requirements that cannot be validated are really just "wishes." Therefore, an important task entails planning how to verify each requirement. In most cases, designing acceptance tests accomplishes this. Identifying and designing acceptance tests may be difficult for non-functional requirements to be validated. They must first be analyzed to the point where they can be expressed quantitatively [SWEBOK 2004].

6.2 Software verification and validation (V&V).

As mentioned earlier, requirements (documents) should be subject to V&V procedures. These are independent procedures, which are used together for checking that a product, service or system meets requirements and specifications and that it fulfills its intended purpose. As previously noted, the words "verification" and "validation" are sometimes preceded with "independent" (or IV&V), indicating that the V&V is to be performed by an impartial third party.

(1) *Verification* — The *evaluation* of whether or not a product, service or system complies with a regulation, requirement, specification, or imposed condition. It is often an internal process. Contrasts with *validation*.

(2) *Validation* — The *assurance* that a product, service or system meets the needs of the customer and other identified stakeholders. It often involves acceptance and suitability with external customers. Contrasts with *verification*.

Verification is intended to check that a product, service or system (or portion thereof, or set thereof) typically meets a set of design specifications. In the development phase, verification procedures involve performing special tests to model or simulate a portion or the entirety of a product, service or system, and then performing a review or analysis of the modeling results.

Validation is intended to ensure that a product, service or system (or portion thereof, or set thereof) results in meeting the operational needs of the user. For a new development flow or verification flow, validation procedures may involve modeling either flow and using simulations to predict faults or gaps that might lead to invalid or incomplete verification or development of a product, service or system (or portion thereof, or set thereof).

To strengthen V&V, there are organizations and corporations that perform a service called "*independent verification and validation*" (IV&V). These organizations or corporations are employed to perform V&V on selected software projects. IV&V is generally very thorough since the industry livelihood depends on finding errors in the software product or process.

It is sometimes said that validation can be expressed by the query, "Are you building the right thing?" and verification by, "Are you building it right?" [Boehm 1981] In practice, the usage of these terms varies. Sometimes they are even used interchangeably. [This description of V&V is credited to Dr. Barry Boehm.]

Requirements may be validated to ensure that the software engineer has understood the requirements. Validation is also important to verify that a requirements document conforms to company standards and that it is understandable, consistent and complete.

It is normal to explicitly schedule one or more points in the requirements process where the requirements are to be validated. The aim is to detect any problems before resources are committed to addressing the requirements. Requirements validation is concerned with the process of examining the requirements document to ensure that it defines the right software (that is, the software that the users expect) [SWEBOK 2004].

Formal notations offer an important advantage by permitting the last two properties to be proven (in a restricted sense, at least). Different stakeholders, including representatives of the customer and developer, should review the document (see the prior discussion on the use of a users' manual as a means of verifying the software requirements specification) [SWEBOK 2004].

6.3 Requirements reviews.

Perhaps the most common means of validation is by *inspection* or *review* of the requirements documents.

 group of reviewers is assigned to look for errors, mistaken assumptions, lack of clarity, and deviation from standard practice. The composition of the group that conducts the review is important (at least one representative of the customer should be included for a customer-driven project, for example), and it may help to provide guidance outlining what to look for in the form of checklists.

Reviews may be imitated on completion of the system definition document, system specification document, SRS document, baseline specification for a new release, or at any other step in the process [SWEBOK 2004].

6.4 Requirements attributes.

Requirements should consist not only of a specification of what is required, but also of ancillary information that helps to manage and interpret the requirements. This should include the various classification dimensions of the requirement and the verification method or acceptance test plan. It may also include additional information such as a summary rationale for each requirement, the source of each requirement and a change history. However, the most important requirements attribute is an identifier that allows the requirements to be uniquely and unambiguously identified [SWEBOK 2004].

Essential properties (attributes) of a software requirement for all software requirements specifications are listed in Table 2.

6.5 Requirements tracing.

Requirements tracing is concerned with recovering the source of requirements and predicting their effects. Tracing is fundamental to performing impact analysis when requirements change. A requirement should be traceable to the requirements and stakeholders that motivated it (from a software requirement back to the system requirements that it helps to satisfy, for example). (*See Figure 6 for an illustration of requirements tracing.*)

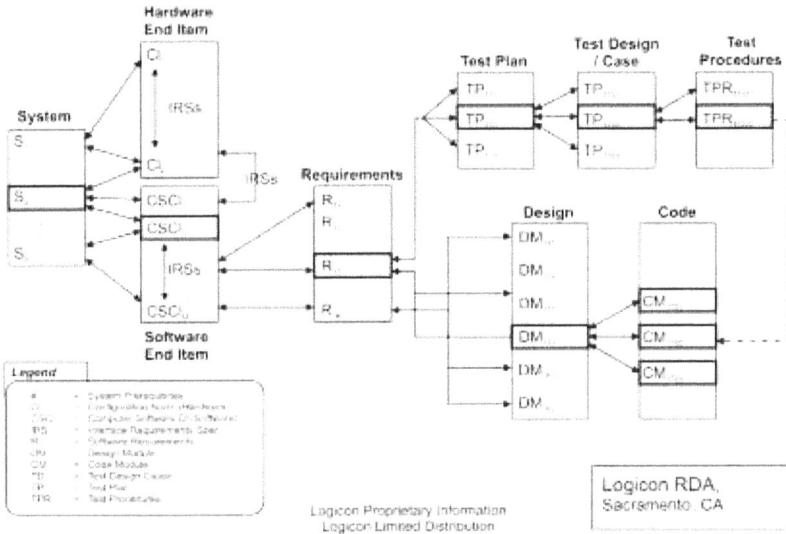

Figure 6: Requirements tracing

6.6 Prototyping.

Software prototyping is the activity of creating prototypes of software applications, i.e., incomplete versions of the software program being developed. It is an activity that can occur in software development and is comparable to prototyping as performed in other fields, such as mechanical engineering or manufacturing [https://en.wikipedia.org/wiki/Prototype].

A prototype typically simulates only a few aspects of, and may be completely different from, the final product.

Prototyping has several benefits: The software designer and implementer can obtain valuable feedback from the users early in the development phase. The client and the contractor can compare if the software design matches the software specification according to which the software program is built.

Prototyping also allows the software engineer some insight into the accuracy of initial project estimates and whether the proposed deadlines and milestones can be successfully met. The degree of completeness and the techniques used in prototyping have been in development and debate since they were proposed in the early 1970s [https://en.wikipe dia.org/wiki/Software_prototyping].

Prototyping is both a front-end and a back-end tool. It is commonly a means for validating the software engineer's interpretation of the software requirements as well as for eliciting new requirements. The advantage of prototypes is that it is easier to interpret the software engineer's assumptions, and where needed, provide useful feedback examining why these assumptions are wrong. For example, the dynamic behavior of a user interface can be better understood through an animated prototype than through textual description or graphical models.

However, there are also disadvantages to the use of prototypes. These include the danger of users' attention being distracted from the core underlying functionality by cosmetic issues or quality problems with the prototype [SWEBOK 2004]. It is also important that *everyone understands that a prototype is not a deliverable system*. Prototypes, by definition, are draft systems without the usual "bells and whistles" of a "full-blown" software system (i.e., a system not completely tested, that may not have a user's manual, perhaps containing outdated display monitors, and definitely not maintainable). Prototypes should not be delivered to the customers, no matter how hard they plead.

> *Note: I have experience delivering a prototype because my boss "overrode" my decision not to deliver the prototype to the "customer." ('After all, they had paid for it', he said). As soon as the prototype was received, this customer began a series of complaints citing how poorly the system quality was designed and insisted that we fix the errors. (These were not really errors but areas we did not feel were necessary to prototype.) I left the company soon after that situation.*

6.7 Model validation.

It is typically necessary to validate the quality of the models developed during analysis. For example, when validating object models, it is useful to perform a static analysis to verify that communication paths exist between objects. If formal specification notations are used, it is possible to use formal reasoning to prove specification properties [SWEBOK 2004].

7. Requirements Management

7.1 Process management.

This topic introduces the project management resources required and consumed by the requirements process.

Two types of management exist in a SWE project. These are (1) *project management* that involves *planning, organizing, staffing, directing*, and *controlling* the project, and (2) *technical management*, or *system engineering*, that involves *problem definition, solution analysis, project planning, process control*, and *product evaluation.*

Major issues in software requirements management include:

(1) It is very important to recognize that a successful manager must be able to develop a system with requirements that are not clear.

(2) The manager must also be able to maintain goals and objectives with projects containing shifting requirements.

(3) It is important that the project manager stick to the principles of SER even when opposed by those stakeholders that do not believe that SER is cost effective.

7.2 Change management.

Change management is central to the management of requirements. This topic describes the role of change management, the procedures that need to be in place and the analysis that should be applied to proposed changes. Change management typically refers to *software configuration management.*

7.3 Project management responsibilities.

Project management can be separated into five components: *planning, organizing, staffing, directing,* and *controlling.* All the management activities such as budgeting, scheduling, establishing authority and responsibility, relationships, training, communicating, and so forth fall under one of these five headings. It should also be noted that management performs many non-management functions (e.g., personnel training, eating a brown-bag lunch, obtaining security clearances, and attending conferences) that are not included in these responsibilities.

7.4 Technical management responsibilities.

Technical management is responsible for the adequacy of the requirements specifications. A major tool of technical management is software configuration management. The technical manager needs to control the volatility of requirements, manage change history, perform requirements traceability, and negotiate requirements changes between the acquirer (customer) and the developer.

7.5 Measuring requirements.

As a practical matter, it is typically useful to have concept of the volume of requirements necessary to manage a particular software product. This volume is useful in evaluating the size of a proposed change to requirements, in estimating the cost of a development or maintenance task, or simply in using the volume as the denominator of other measurements. Functional size measurement (FSM) is a technique is used to evaluate the size of a body of functional requirements. IEEE Standard 14143.1 defines the concept of FSM.

8. Requirements Tools

Tools for dealing with software requirements fall broadly into two categories: tools for modeling and/or identifying requirements and tools for managing requirements activities.

8.1 Software engineering requirements tools.

Some of the software engineering requirements tools available to project managers are discussed below. It should be noted that a tool is not necessarily a mechanical or automated device that provides an extension to the software engineers' human capabilities (for example, one of the conceptual models listed in Paragraph 2 earlier). A tool can be a process, an approach, a development standard, and/or even a person.

Requirements modeling tools are tools used for eliciting, recording, analyzing, and validating software requirements. (*See Paragraph 4.2 discussing Requirements analysis.*)

8.2 Software requirements management tools.

Requirements management tools primarily involve tracing and prototyping tools. *Tracing tools* are tools used for the identification and documentation of the derivation path (upward) and allocation/flowdown path (downward) of requirements in the requirements hierarchy. [SWEBOK 2004] A *prototyping tool* uses an early sample, model or release of a product built to test a concept or process or to act as an object to be replicated or learned.

Other requirements management tools support a range of activities—including documentation, change management, risk management, and so forth—and have had significant impacts on design. These same tools are used to support the primary duties of software project management.

In practice, many organizations have invested in requirements management tools, although others manage their requirements in ad hoc and generally less satisfactory ways (e.g., using spreadsheets). The following five tools are identified as supporting software engineering requirements.

(1) **Project planning** — Work breakdown structure (WBS), critical path method (CPM), PERT charts, Gantt charts, etc.

(2) **Cost analysis tools** — *Algorithmic models* (algorithms for producing a software cost estimate—COCOMO is an example), *rules of thumb* (guidelines that have evolved over time), and *estimation by analogy* (compared to completed projects).

(3) **Project tracking and control tools** — *Baseline management* (partitioning the project into manageable phases), *configuration management* (controlling the project documents) and *quality* (process) *assurance.*

(4) **Risk management tools** — Critical path method (CPM), top-ten track charts or contingency plans.

(5) **Communications tools** — The act of managing a project with the assistance or support of a data processing system. For example, word processing, outliners, notepads, auto dialer, appointment schedulers, electronic mail, etc.

8.3 Software technical manager tools.

Technical managers, who are also called *system engineers* or *technical leads*, have a different job to perform, and as a result, complete tasks using various tools. Indeed, tracing and change management are only practical if supported by a tool.

(1) **Work breakdown structures (WBS)** — A tool for representing relationships.

(2) **Functional flow block diagrams (FFBD)** —A tool for "roughing out" the major functions and their interfaces.

(3) **Dataflow diagrams** — A tool for representing dependencies.

(4) **N-squared (N^2) charts** — An aid to establishing and representing system and subsystem interfaces.

(5) **Timeline analysis** — Depicts the concurrence, overlap and sequential relationship of time-critical functions and related tasks.

(6) **Test compliance matrix** — A representation method for displaying what methods will be used to verify ("test") each software requirement.

References

- **[Boehm 1981]** Barry W. Boehm, *Software Engineering Economics*, Prentice Hall, Upper Saddle River, NJ, 1981.

- **[Fairley & Thayer 1997]** Richard E. Fairley and Richard Hall Thayer, "The Concept of Operation: The Bridge from Operational Requirements to Technical Specifications," in *Annals of Software Engineering*, Vol. 3, issue 1, 1997, pp. 417–432.

- **[IEEE Std. 610.12-1990]** *IEEE Standard Glossary of Software Engineering Terminology,* IEEE, Inc., New York, 1998.

- **IEEE Std. 830-1998]** *IEEE Recommended Practice for Software Requirements*, IEEE, Inc., New York, 1998.

- **[IEEE Std. 12207.1-1997]** IEEE Implementation of International Standard ISO/IEC 12207:1995, Standard for Information Technology - Software Life Cycle Processes - Life Cycle Data.

- **[Justis & Kreigsmann 1979]** R.T. Justis and B. Kreigsmann, "The Feasibility Study as a Tool for Venture Analysis," *Business Journal of Small Business Management*, 17(1), 1979, pp. 35-42.

- **[RADC TR-83-175]** Bowen, T.P., J.V. Post, J.T. Tsai, P.E. Presson, and R.L. Schmidt, *Software Quality Measurement for Distributed Systems*: Vol. 1, *Final Report*; Vol. 2, *Guidebook for Software Quality Measurement*; Vol. 3, *Distributed Computing Systems: Impact on Software Quality*, RADC TR-83-175, prepared by Boeing Aerospace Company for Rome Air Development Center, Griffiss AFB, NY, 1983.

- **[RADC TR-85-37]** Bowen, T.P., G.B. Wigle, and J.T. Tsai, *Specification of Software Quality Attributes*: Vol. 1, *Final Technical Report*; Vol. 2, *Software Quality Specifications Guidebook*; Vol. 3, *Software Quality Evaluation Guidebook*; RADC TR-85-37, prepared by Boeing Aerospace Company for Rome Air Development Center, Griffiss AFB, NY, February 1985.

- **[Sommerville 2007]** Ian Sommerville, *Software Engineering*. Addison-Wesley, Boston, 2007.

- **[Sommerville 2010]** Ian Sommerville, *Software Engineering*. Addison-Wesley, Boston, 2010.

- **[SWEBOK 2004]** *Guide to the SWE Body of Knowledge*, IEEE, New York, 2004.

- **[SWEBOK 2014]** *Guide to the SWE Body of Knowledge*, IEEE, New York, 2014.

- **[Wasserman and Pircher 1987]** A.I. Wasserman and P.A. Pircher, "A Graphical, Extensible, Integrated Environment for Software Development," *SigPlan Notices*, Vol. 22, no. 1, January 1997, pp. 131-142.

- **[Yourdon 1989]** Edward Yourdon, *Modern Structured Analysis*, Prentice-Hall, Upper Saddle River, NJ, 1989.

Chapter 2
Software Requirements[3]

Peter Sawyer
Lancaster University, U.K.

Abstract

Software requirements concern the specifications of software systems. Software requirements are always derived from a business problem, for example, processing passport applications, improving automotive safety systems or adding features to cell phones. Sometimes the solution will be implemented entirely using software. At other times, software will be only one of the technologies that needs to be integrated into the solution. Projects may develop new software products or they may be concerned with evolving existing systems. In all cases, the software will be embedded in an operational context, and therefore, will have interfaces to human users, business process elements or other software or hardware systems (Figure 1).

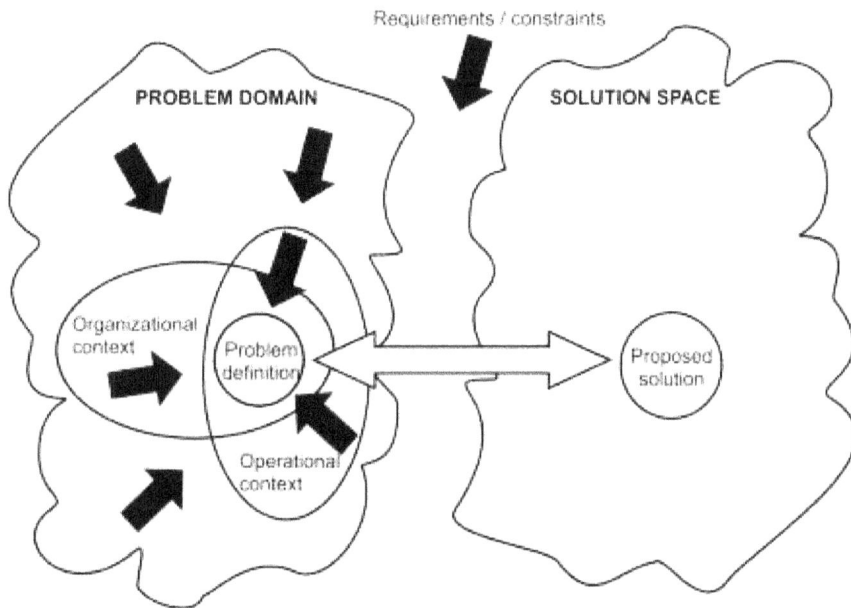

Figure 1: Requirements, constraints, problems, and solutions in requirements engineering

---

3. This document originally appeared in R.H. Thayer and M. Dorfman, *Software Engineering: The Development Process,* IEEE Computer Society Press, Los Alamitos, CA, 2005. Used with permission of the author.

1. The Contents of Software Requirements

The job of developing software requirements can very rarely be isolated from the underlying business problem. In most cases, the software requirements have to be developed with full awareness of the real world context. Hence, software requirements are not really a discrete area of software engineering. Rather they are part of the systems engineering process known as requirements engineering (RE). RE will deal with all aspects of problem understanding and solution specification.

Transforming a requirement into software is a complex process that consumes many resources. The deeper into the process, the more design and implementation strategies become committed to satisfying the requirement. The effect of this is to make the costs of rectifying errors in the requirements increase dramatically as development proceeds. An effective RE process which minimizes the occurrence of requirements errors is therefore critical to the success of any development project.

The RE process is the responsibility of the *requirements engineer.* Requirements analyst, systems analyst and business analyst are common alternative names for requirements engineer. The term *business analyst* is particularly revealing because it correctly implies that the requirements engineer's role is not purely a technical one. A business analyst must be able to understand the domain of the business problem and translate the problem into a specification for a software solution.

Hence, the requirements engineer bridges the business and technical domains. This clearly calls for a combination of skills. A software engineer performing the role of a requirements engineer (requirements engineers do not always have a software background) needs to supplement their technical skills and knowledge with an understading of business domains and the "people" skills necessary to draw information from customers, users and other stakeholders.

2. Requirements and Constraints

A *requirement* defines a property or capability that must be exhibited by a system in order for it to solve the business problem for which it was conceived. Thousands of requirements, of different kinds and at different levels of abstraction, may be needed for a challenging business problem.

The classic way to categorize requirements is according to whether they are functional or non-functional. A *functional requirement* describes a function that the software must perform. For example, a requirement for a burglar alarm system to *activate the siren when a sensor is tripped* is a functional requirement. It defines an action to be performed by the burglar alarm on the occurrence of some stimulus, like a burglar trying to enter a house.

Non-functional requirements describe qualities of a system. Common types of non-functional requirements include reliability, availability, security, safety, usability, and performance requirements.

The most important class of non-functional requirements should address how well the system operates within the environment and the quality of the requirement—sometimes called "extra-functional requirements [4].

Some requirements are *emergent properties*. These are dependent on a wide range of factors, some of which are hard to analyze and control. Requirements engineers and developers feel most comfortable when a requirement for a system property can confidently be allocated to a particular system component that assumes responsibility for satisfying the requirement. Emergent properties are the antithesis of this where satisfying a requirement depends upon how the whole system operates within its environment. For example, the performance of a distributed system may be an emergent property because it is dependent upon the design of the system but also on factors such as demand and network traffic.

Requirements are usually specified at several points on a spectrum ranging from those with a business focus to those with a technical focus. The technically focused requirements exist only to make it possible to satisfy the business-focused requirements. For convenience, we will treat these as levels where the highest-level requirements are those with a business focus.

At the highest level are the goals of the system that set out in very broad strategic terms what is needed to solve a defined business problem. Identifying these goals is what typically motivates a development project. The next level of requirements define what must be observable (by the people or other systems in the environment) of a black box system that would solve the business problem. These are often called the user requirements. This term is a little unsatisfactory, however, because users, the people who will interact directly with the system, are not the only sources of requirements.

Stakeholder requirements are more complete because they are inclusive of users and others who impose important requirements, such as managers or external regulators. Requirements may come from non-human sources such as the environment, or the application domain (Figure 1). For this reason, the term *system requirements* is used throughout this chapter to describe required properties observable by users, other stakeholders and other systems in the system's environment.

One of the main tasks of requirements analysis is to elaborate the system requirements to discover more about the implications of satisfying them. This involves deriving new, lower level requirements (called *derived requirements)* that focus more on detailed technical issues (confusingly, technically-focused

---

4. Another class of nonfunctional requirements addresses nonoperational issues such as maintainability.

requirements are sometimes called system requirements, although not in this document).

Software requirements are the requirements that a software system or component has the task of satisfying. In a domestic burglar alarm system, only a subset of the requirements would be software requirements because other requirements would specify the properties of hardware components such as sensors and sirens. A *personal computer* (PC) image manipulation application, by contrast, would run in a software environment and involve no hardware components. Even here, not all the requirements may be software requirements because there may also be requirements for an installation guide and a user manual.

Constraints are like negative requirements. They act to limit the set of possible solutions to the business problem. For example, noise legislation may act as a constraint that limits the length of time that a burglar alarm is permitted to sound its siren. Many constraints are technical and concern, for example, memory, processor speed and bandwidth. Others are related to the problem domain. For example, on-board train control systems are constrained by high levels of electro-magnetic radiation in the trackside and on-board environments.

3. Requirements Engineering Process

A *requirements engineering* (RE) process must transform a business problem into a specification of the properties of a system that will provide an appropriate solution to the problem. Arriving at a solution specification requires the application of a systematic and rigorous process in order to understand the problem and the impact of a range of possible solutions.

The RE process is broadly composed of the set of activities reflected in the structure of the following sections of this chapter. The properties that the software must exhibit have to be elicited. The elicited requirements need to be subjected to analysis.

In order to establish a set of requirements that is correct, complete and feasible, the set of requirements that emerge from the analysis activity need to be recorded in a specification document that communicates the requirements to the people who will use them to develop the software. The documented requirements need to be validated to ensure that the software that they specify will meet the needs of the people from whom the requirements were elicited (Figure 2). As development proceeds, the requirements need to be managed so that changes are controlled.

RE is never a strictly linear process. For example, analysis of the requirements is closely coupled invariably with their elicitation so that the requirements engineer and the stakeholders can work closely to develop a consistent understanding of the system requirements.

The IEEE publishes three standards that directly address the RE process and that have evolved to make the handling of requirements rigorous and controllable.

1. IEEE Std. 1362-1998 *Guide for Information Technology - System Definition - Concept of Operations (ConOps)* Document [IEEE 1998a]

2. IEEE Std. 1233-1998 *Guide for Developing System Requirements Specifications* [IEEE 1998b]

3. IEEE Std. 830-1998 *Recommended Practice for Software Requirements Specifications* [IEEE 1998c]

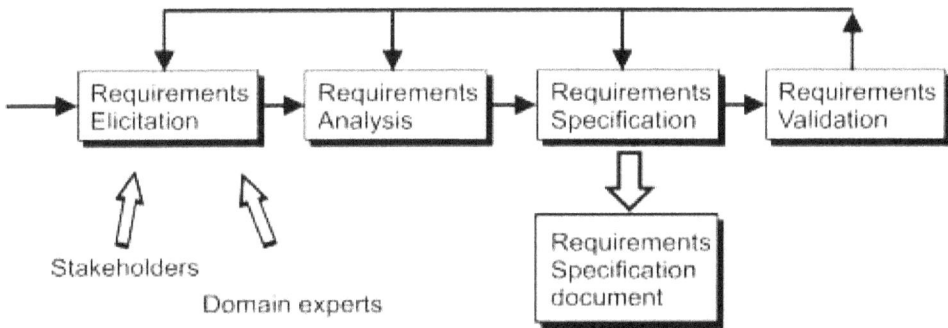

Figure 2: Generic model of the RE process

The process implicit in this set of standards begins with scoping the system. This involves understanding the underlying problem that the system is to address, identifying the goals of the system and outlining how the system will operate in its environment. This is the concept of operations or *ConOps*. This is followed by a process in which the system requirements are elicited from their sources, analyzed and validated. The product of this is a *system requirements specification* document that defines the requirements for the overall system.

For each software component, further analysis of the allocated requirements is used to derive the requirements that fully specify the software. These are documented in a Software Requirements Specification (SRS). The SRS forms the definitive set of requirements that the component must satisfy and is detailed sufficiently to allow development to commence. The development of a system architecture from the *system requirements specification* is an implicit prerequisite for the identification of the software (and other) components.

This process has evolved to deal with large systems engineering projects typically comprising both hardware and software. Such systems are often developed using a supply chain of subcontractors who are responsible for delivering system components and managing a hierarchical relationship with a main contractor. The complexity of such projects means that very serious risks are posed by requirements change (e.g., the emergence of new, erroneous or poorly understood requirements) late in the development process. This complexity needs an RE process which seeks above all to minimize the risk of unexpected requirements change. The process achieves this by ensuring that the problem and its solution are rigorously analyzed and documented before serious development begins. Some requirements changes will still occur, but the extent and impact of the changes should be controllable.

In some domains, it is impossible to identify all the requirements that will serve to define a product for its expected service life. If the product's environment is volatile, the product's requirements will also be volatile. Where software products are developed to compete in a competitive open market, for example, the requirements are likely to evolve with the market. Other pressures may also influence the RE process. In particular, capturing market share sometimes means that meeting release date deadlines is traded off against rigor in the RE process. This trade-off is a delicate one that can easily result in short-term expediency, storing problems that will appear later in the product life cycle if, for example, requirements are poorly managed.

Despite this caveat, small projects to develop non-critical software undertaken by companies with relatively simple organizational and management structures may be able to cope with some requirements changes providing the likelihood and extent of change is predicted and acknowledged. Such projects need mechanisms in place to enable them to adapt gracefully when requirements change occurs.

One increasingly common response to coping with change is to design the development process for frequent releases of the software where development proceeds in relatively small increments. The RE process needs to be synchronized with this by being enacted in increments (Figure 3). Here, requirements to be implemented in each release are selected from a pool of requirements held over from the previous release, and of newly emerged requirements.

The *incremental RE process* depicted in Figure 3 proceeds in parallel with other development phases. When release *n* of the software is being implemented, development of release *n+1* is prepared by selecting the requirements that it is to satisfy. It is clear that in this process RE is not simply a front-end activity that ceases when the requirements have been documented and design commences. Instead, a continuous process lasts throughout the entire product life cycle.

In fact, this is true of all RE processes. Even when the process is optimized to minimize requirements change, some reworking of requirements is inevitable after design and coding has commenced. The RE process normally consumes most effort early in a project, but efforts need to be allocated to requirements management and must cope with change following sign-off of the requirements specification.

Only two very broad types of RE processes have been outlined above. In practice, there are many different models of the development process and each needs an RE process that is tailored accordingly. Despite the variety of RE processes and the range of practices used for elicitation, analysis, specification, and management, (*see, for example* [Robertson & Robertson 1999, Sommerville & Sawyer 1997, Wiegers 2003, and Young 2001]) the essential goal of an RE process always applies. This is to derive and specify an appropriate solution to a business problem. To achieve this solution, requirements have to be discovered, understood, recorded, checked, communicated, and managed.

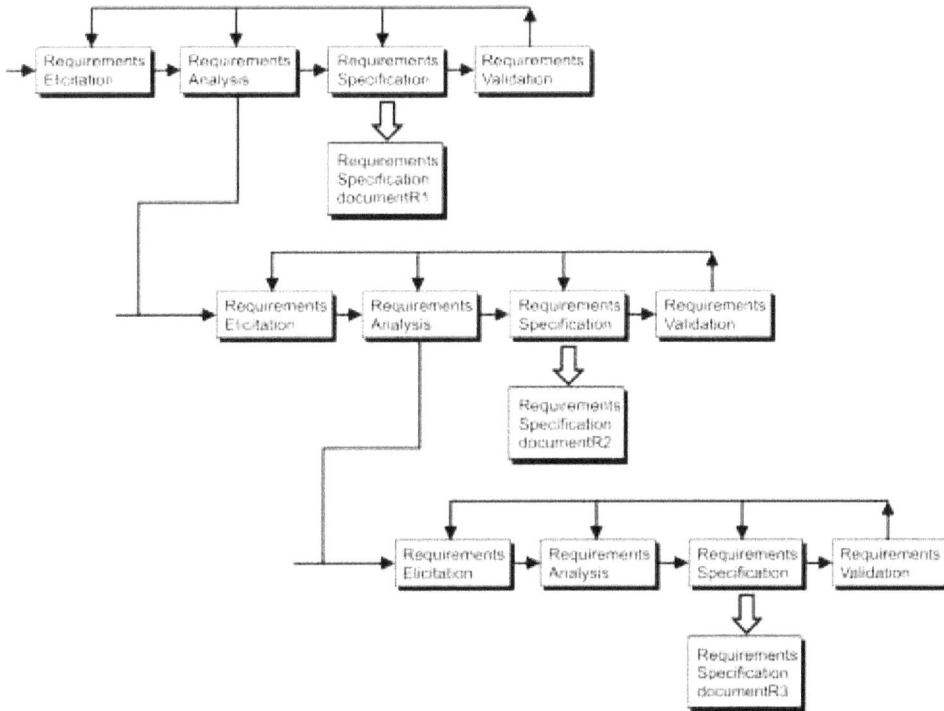

Figure 3: An incremental RE process

4. Requirements Elicitation

Requirements elicitation is the process of discovering the requirements. The requirements engineer needs to identify the sources of the requirements, collect information about the problem from these sources and synthesize requirements from the information. As always, this is not a linear process but one that requires *iteration* as information is collected, clarified, corrected, and reformulated.

However, elicitatiom is not simply about the requirements engineer passively learning what the systems stakeholders require. The requirements engineer needs to dig beneath the stakeholders' accounts of their concerns, needs and desires to uncover the underlying business problem that gives rise to them. Although the requirements engineer needs to learn about the problem, the stakeholders also need to understand what can be achieved and the relationships between their requirements and those of other stakeholders.

The key point is that elicitation—the name routinely applied to this activity—is misleading. Requirements are not simply elicited fully formed from stakeholders. They have to be *discovered*, whether their sources are human, documentary or the environment. Another way to think of this is that requirements that will be documented in the specification are synthesized from the elicited requirements information.

Systems are normally commissioned to address a strategic goal or goals and these requirements are the first things that have to be discovered. The requirements subsequently discovered must contribute to meeting these goals. When the project goals are known, the scope of the project can be defined. This may include management or process issues that are outside the remit of RE. An important element of the scope, however, is defining the system boundary. This is dealt with under requirements analysis. From the elicitation point of view, an explicit definition of the project scope helps to ensure that elicitation of the requirements is focused on relevant issues.

4.1 Requirements sources.

A large system will have many different stakeholders and in most cases, these will be the primary sources of the requirements. Indeed, any project that does not engage with the people who will actually pay for and use the proposed system is bound to fail.

Identifying the *stakeholders* is a crucial first step. The type of project is a factor in the identification process. A project, which is developing a custom system for a commissioning customer, will more easily identify representative stakeholders than a project developing a product to sell to a market. Stakeholders need to be classified in order to ensure that no significant sources of requirements are overlooked. The role of each stakeholder needs to be understood and representatives of each type of stakeholder must be selected to work with the requirements engineer. Most stakeholders are busy people so it is vital to find those who are motivated to act as *product champions* [Wiegers 2003] and to negotiate access to them.

Stakeholders have viewpoints. These are partial views of the problem domain that are colored by the stakeholders' own roles and experience. Inevitably, this means that their requirements may not be consistent with those of other stakeholders. Recognizing the scope and limitations of stakeholders' viewpoints is important. It can help the requirements engineer to impose a separation of concerns on the set of requirements and therefore, to help resolve inconsistencies and apply priorities.

Stakeholders do not represent the only sources of requirements. Embedded software may have no direct users, for example. However, the system in which the software is embedded will have users and other stakeholders and the embedded software requirements will be derived from these.

Key requirements and constraints often come from the application domain (burglar alarm systems may be subject to noise legislation, for example) and from the business rules that exist in the organizational environment (for defining usage privileges, for example). Key requirements may therefore be hidden in documents and in interface specifications and held within the experience of domain experts.

Domain expertise plays a crucial role in successful requirements elicitation. It is needed to identify the key requirements and constraints that are derived from the application domain. Although stakeholders have domain expertise, they are not ideal choices to play the role of domain expert. This is partly because of their viewpoints but also because they often find it difficult to articulate the tacit knowledge that often underpins how the domain operates. The domain expert's role is to help make this tacit knowledge explicit.

The requirements engineer may have sufficient domain expertise if they have previous experience in the application domain. However, in an unfamiliar, complex domain, it may be infeasible for the requirements engineer to acquire domain expertise during the course of the project, even with training. In this case, domain expertise may have to be brought in from elsewhere to ensure an active role in the RE process.

4.2 Elicitation techniques.

Once the requirements sources have been identified, the process of collecting knowledge about the problem—the raw material from which requirements will be synthesized—can begin. This starts with understanding the problem. Only once the problem is understood will the *elicited requirements* start to make sense. A key factor in understanding the problem is the stakeholders' role in the problem domain. In most cases, this means understanding their jobs.

Many jobs involve subtle facets and the interaction of people with different roles and expertise. Failure to identify these can result in a system that fails.

However, capturing the subtleties can be difficult because people often find it challenging to articulate all the nuances of their job. Because of this, simply asking a user "what do you need from the system?" will not work. The requirements engineer first needs to understand the problem and then find an effective way to get the stakeholders to analyze what is needed. The requirements engineer will then be better able to reconcile any inconsistencies and assess what is needed and is feasible. In order to do this, the requirements engineer needs to find a means to tease out the information in a way that gives stakeholders some context within which to describe their role.

For sociotechnical systems, users' stories or scenarios can provide a valuable tool. Here, the requirements engineer convinces users to identify their principle tasks. Each task is described as a sequence of events noting the preconditions, postconditions, communications with colleagues, and other events that comprise the task. Use cases mesh well with scenarios. Here, use cases are simply top-level user tasks within the problem domain that are described in terms of scenarios.

Using scenarios helps the user to articulate their tasks by requiring them to describe the constituent activities. They help the requirements engineer by getting the user to focus on the essential characteristics of their task. The requirements engineer can better understand how jobs are performed and how they are organized within the work environment. Scenarios provide cues that allow the requirements engineer to ask for clarification at key points ("what happens if *<resource x>* is unavailable when you need it?").

Once the problem is understood, use cases can be used again to envision ways of interacting with the proposed system. Here, the use cases represent the top-level services offered by the system. Scenarios are then developed by the stakeholders and the requirements engineer for each use case to explore how it should operate, the interactions necessary to make it operate, and the exceptions to the normal sequence of events that have to be handled. The requirements implicit in the use case scenarios must then be made explicit.

Elicitation can proceed as a series of interviews between the requirements engineer and individual stakeholders but it is often useful to get the stakeholders together in one place. This is particularly useful once the requirements engineer has progressed beyond understanding the problem and has begun to elicit the requirements of the proposed system. Stakeholders' restricted viewpoints may lead them to propose requirements that are hard to reconcile with those of other stakeholders.

For example, there may simply be too few resources available in the project to satisfy everyone's requirements. If run skillfully, elicitation workshops can provide an efficient way of resolving such problems by exposing stakeholders to others' ideas and concerns and by fostering a sense of collective responsibility for their resolution. Care must be taken to ensure fairness and counter phenomena such as *groupthink.*

However, the use of scenarios and workshops does not preclude the use of other techniques and may not be the best techniques for every project. Some stakeholders may be unwilling or unable to participate in workshops or may find scenarios an awkward mechanism for articulating their requirements. Senior managers, for example, may not interact directly with a system and so may find it hard to talk about the system in terms of scenarios.

Traditional, individual interviews, in which the requirements engineer gets the stakeholder to state directly the properties and qualities required of the proposed system, may be necessary. Even here, however, the requirements engineer needs to understand (at least) the rationale for each requirement and the importance that the stakeholder attaches to it. Observation of users in their working environment may also provide revealing insights about how the work is actually done. For example, the physical layout of a control room may influence working practices in ways that can only be appreciated when directly observed.

The techniques discussed above emphasize the role of system users. Sometimes, however, access to users may be hard. When a product is being developed for a market rather than for an identified customer it may be hard to identify representative users. At the very least, a user's champion must be identified. Competing products and emerging trends and technologies are also likely to be important sources of requirements.

Special attention must be given to discovering non-functional requirements. One important subset of non-functional requirements is that concerned with dependability. If a high degree of dependability is needed, considerable resources need to be devoted to the discovery and analysis of the dependability requirements. In safety-critical domains, a hazard analysis can be used to identify the hazards and derive safety requirements to counter them. In systems with serious security implications, software security experts are needed to assess the security vulnerabilities and derive requirements.

A specialist requirements practice of the kind adopted for dependable systems is expensive and is economically impractical for most other systems. Non-functional requirements may nevertheless be very important for these systems. Stakeholders and domain experts will usually have an appreciation of the critical quality aspects of the problem domain. However, the requirements engineer must recognize the required quality characteristics of the proposed solution. For example, manual business processes are often fault-tolerant because of the natural adaptability of the people who perform them.

Automation risks reduce dependability unless the appropriate reliability, availability or other non-functional requirements are recognized and specified. A deep understanding of the implications of the proposed system is needed if the requirements engineer is to recognize such requirements.

5. Requirements Analysis

Requirements analysis pertains to understanding the problem and synthesizing a set of requirements that specify the best solution. Analysis is needed to help deepen understanding of the problem and what is required, and to detect and resolve problems such as inconsistencies and incompatibilities with the requirements.

Elicitation and analysis are closely coupled activities. Requirements information has to be analyzed as it is elicited in order to understand, for example, what question to ask next. Among the questions to ask are *why* questions—why is it done this way? Deriving the best solution may involve challenging stakeholders' perceptions and methods of doing things.

Ultimately, the analysis activity should yield a *baseline* set of requirements. This is the set of requirements that the system (or version of the system) will implement. Resource limitations often mean that the requirements baseline specifies a subset of the features requested by the stakeholders. The requirements engineer and the stakeholders have to agree on what is to be included and what is to be left out. Inconsistencies and conflicting requirements must be negotiated and traded-off. The requirements in the final product should be both necessary and sufficient.

This means that the baseline should not include any requirements that do not contribute to the goals of the project. It must be complete—there must be no requirements missing that are necessary to specify how the problem is to be solved.

5.1 The system boundary.

The first analysis activity takes place when the system goals are defined (*see Paragraph 4*)—the project needs to be scoped. The scope of the project may include things to do with, for example, the development process or procurement policies. However, the scope must also include a definition of the *system boundary.* This is concerned specifically with identifying which elements of the problem are to be addressed by the proposed system.

The system boundary should ensure that the proposed system focuses on satisfying the project goals. Components within the system's environment that may impose requirements on or constrain the system are placed outside the system boundary. Within the system boundary are many of the aspects of the problem for which the proposed system will provide a solution.

For example, a project to develop a domestic burglar alarm system containing all the functions concerned with protecting the house from intruders needs to be placed inside the system boundary. Outside the system boundary will be other systems in the alarm's environment that interface to the alarm. For example, the alarm might be connected via a communications link to a security service.

The functions needed to permit the security service to monitor the alarm would be placed inside the system boundary, so requirements for these need to be discovered and specified in the system specification. However, specifying the security service's reaction to a tripped alarm would be out of scope for the system requirements and so are placed outside the boundary. There are a number of ways to depict the system boundary. One way is to apply a use case diagram. Figure 4 depicts the system boundary for a burglar alarm system using a use case.

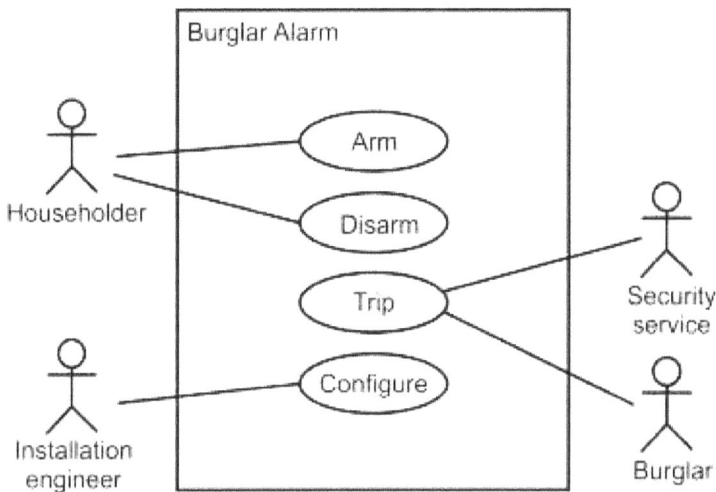

Figure 4: Use case defining the system boundaries for a domestic burglar alarm

Defining the system boundary involves envisioning how the proposed system will operate in its environment. This is one of the main advantages of explicitly developing a concept of operations [IEEE 1998a], which involves elicitation of the project goals, analysis of possible solutions and definition of the system boundary. The concept of operations helps to verify the project's viability and ensures that project participants understand what they are trying to achieve before commencing the main elicitation and analysis activities.

5.2 Requirements modeling.

In any RE project, there is a considerable amount to understand. To help make sense of complex information, engineers use *models*. Requirements engineers construct models of the problem to discover suitable solution requirements. Models are also used to help describe the proposed system to help communicate requirements to the developers.

There are many different modelling notations. In general terms, the same notations are useful for both problem and solution modelling, for example, an object model (or a class diagram in UML) [Rumbaugh, Jacobson, & Booch 1999] may be constructed to understand the structure of a business activity just as an object model may be used to define the structure of a system that automates the activity.

Models help to explore the requirements by providing alternative depictions of the problem to supplement those provided by verbal or textual descriptions. Constructing a model helps the requirements engineer to make sense of the information as it is transformed using the rules of the chosen modelling notation. Graphical models help engineers visualize complex system properties using appropriate abstractions. For example, a UML sequence diagram makes explicit the sequence events expected to occur and the interactions between people or system components in order to perform some task. While such models encourage rigor in the analysis they do not enforce it. Formal specification notations such as Z [Woodcock & Davies 1996] or CSP [Schneider 1999] (or even UML's OCL) do enforce analytical rigor.

Formal models are traceable to formal reasoning and enable some properties of system models to be proven. Sometimes the dynamic behavior of systems cannot be adequately analyzed using static models. In these circumstances, it may be necessary to conduct simulations. If uncertainties about systems' dynamic behavior affect its feasibility then simulations must be conducted early in the requirements process.

5.3 Derived requirements.

The cost and technical implications of system requirements are often unclear and this makes them difficult to assess and validate. To make the implications clear, it is usually helpful to elaborate the system requirements by *deriving* new requirements that focus on finer-grained properties of the proposed system. Modelling helps uncover the details of the derived requirements.

In the burglar alarm example, the use case *arm* responds to the system requirement that the householder must be able to arm the alarm when they exit the premises. If the control panel is located inside the premises, then the requirements engineer needs to derive the additional requirements that are an implementation of this. In this example, a derived requirement needs to indicate a time delay between the user instigating the arming process and the alarm becoming armed. In a use case model, such a derived requirement would naturally emerge from writing the following *arm* scenario(s):

5.4 "Arm" primary scenario.

The organizational or operational environments may also yield derived requirements. Knowledge of the domestic (operational) environment may derive a requirement that the alarm must be protected against being armed accidentally by, for example, the family dog brushing against the control panel.

Precondition: Alarm is unarmed.

(1)   *Householder initiates arming procedure.*

(2)   *Timer countdown commences.*

(3)   *Householder exits premises.*

(4)   *Timer countdown finishes.*

(5)   *Arming procedure completes.*

Post condition: Alarm is armed.

Of course, this is a very simple example and the synthesis of derived requirements is seldom so easy.

Understanding the implications of system requirements is one motivation for deriving requirements that are more detailed. The other reason is to add specificity for the developers. The requirements elicited from the stakeholders will typically be expressed in terms of the application domain. If the application domain is life insurance, for example, then the requirements are likely to be couched in terms of policies, investment vehicles and so forth. Every domain has its own culture and jargon and software developers may not understand these well enough to interpret the requirements correctly.

Essentially, the requirements need to be translated from being domain-centric to being software-centric, from abstract to technical. For example, knowledge about the cost and practicality of mechanical locks might lead to a requirement to use a PIN for arming the alarm being derived from the requirement to guard against errant family pets. This in turn might stimulate analysis to derive the length of PIN required to provide adequate security without being an unreasonable burden on the householder's memory.

The derivation of requirements is not confined to functional requirements. High-level expressions of non-functional requirements need to be explored in more detail. They need to be quantified or transformed into a set of equivalent functional requirements. For example, a requirement that a system be *reliable* is hard to make rational design decisions for and impossible to verify.

The requirements engineer needs to choose a suitable metric (e.g., mean time to failure) and specify how the system must score against this metric. Selection of a suitable value for a non-functional requirement against the selected metric may have a profound effect on the cost of satisfying the requirement and the requirements engineer needs to understand the practical limits needed to achieve high levels of (for example) reliability.

Requirements are not always derived top-down because some requirements emerge from the environment or the domain rather than properties requested by the stakeholders. For example, noise legislation may place a legal limit on the length of time that a burglar alarm siren can sound. Hence, a requirement that the siren must not sound for longer than this period when the alarm is tripped must be included in the specification. Sometimes requirements patterns emerge where an experienced requirements engineer recognizes a problem and is able to identify a known solution without having to iteratively derive and analyze successively more detailed requirements. The process of deriving requirements is driven largely by the engineer's experience.

The derivation process should stop when the requirements are sufficiently specific for the requirements engineer to be confident that the requirements are fully understood, and for the developers to commence solution design. Risks that are too derailed can lead to the creation of premature design decisions that may unnecessarily constrain the design. Inevitably, requirements derivation *does* involve the narrowing of design choice but this narrowing can be justified if the requirements engineer is the person best positioned to make the choice.

This was illustrated above by the derivation of a requirement that specifies the use of a PIN to guard the alarm, and another to specify the length of a PIN. Both of these requirements constrain the design. However, the decision on the use of PINs needs to be informed by several factors that the requirements engineer is better able to assess than the developer who has an exclusively technical focus.

There is an important housekeeping activity related to requirements derivation which is a cornerstone of requirements management. Derived requirements must record the requirements from which they are derived.Over the course of the project, a complete set of derivation relationships must be kept that permits the relationships between system and derived requirements to be traced (*see Paragraph 8, Requirements Management*).

5.5 Requirements attributes.

It is insufficient merely to record the statements of need the requirements express. Requirements have a number of *attributes* that should be assigned values in order to ease their management. The analysis stage is where most of these attribute values are assigned. These attributes may include:

- *Identifier* — Every requirement must be assigned a unique identifier that allows it to be unambiguously referenced.

- *Source* — The source of the requirement may be, for example, a stakeholder from whom it was elicited or a higher-level requirement from which it was derived.

- *Date* — At the time the requirement was formulated.

- *Rationale* — The rationale explains the purpose of the requirement. This helps subsequent analysis, particularly if the requirement is challenged or needs to be reworked at a later stage.

- *Type* — This attribute records, for example, whether the requirement is functional or non-functional, and whether it is a user interface requirement, a safety requirement, etc.

- *Priority* — When there are many requirements and a limited budget, choices must be made about what can be implemented and what has to be left out. The requirements engineer needs to impose some view of priorities on the set of requirements. IEEE Standard 830-1998, for example, recommends classifying requirements as essential, conditional, or optional.

- ***Stability*** — Even if the requirements have been analyzed rigorously, uncertainty may remain about some requirements if e.g., the system's business environment is dynamic. The fact that uncertainty surrounds a requirement should be recorded so that its likely volatility is made explicit and appropriate risk containment measures can be taken.

- ***Verification procedure*** — This attribute defines how to verify that the requirement has been satisfied once the software has been implemented. If the verification procedure is defined at the same time as the requirement, it acts as a useful reality check on the requirements (*see Requirements Validation, Paragraph 7*).

- ***Status*** — The requirement's status records its current position in the life cycle of the requirement (*see Requirements Management, Paragraph 8*).

5.6 Requirements trade-offs.

The stakeholders need to approve the requirements baseline by accepting that some requirements will make the cut but others will not. Analysis will sometimes reveal that requirements from different stakeholders are valid but mutually inconsistent. Even more frequently, there must be sufficient project available to satisfy all the requirements. The stakeholders must be made aware of these conflicts and explicitly involved in making the necessary *trade-offs*.

Agreeing on requirements' priorities helps the trade-off process but it is seldom possible to implement all the high-priority requirements because the cost of implementation has to be balanced against priority. In large projects systematic analytical methods for informing the necessary trade-offs are needed (e.g. [Karlsson & Ryan 1997]). A good discussion of requirements priorities can be found in [Wiegers 2003].

Achieving agreement is often easier if stakeholders are aware of each other's concerns. If the stakeholders have been involved in elicitation workshops, this awareness may already exist and will therefore ease the processes of forming a consensus. If not, workshops specifically aimed at achieving this consensus and deriving an acceptable baseline may be necessary.

6. Software Requirements Specification

The *software requirements specification (SRS)* documents software requirements. Projects may use up to three kinds of "specification" documents at different stages of the RE process. These broadly correspond to the IEEE standards reviewed above which are designed to cope with large, complex projects. A ConOps document explains the project vision and scope, a system specification defines the requirements for the whole system, and an SRS specifies the requirements for a software component or subsystem.

Each document has a different purpose and each addresses successively finer levels of detail. The system specification must demonstrate how the envisaged system will address the problem for which it was conceived. It must be readable by the system stakeholders to enable them to validate the requirements and approve them (i.e., *sign them off*) as the basis for subsequent development.

The SRS, by contrast, is primarily a technical document aimed at developers. Although its focus is on specifying the software, some motivation and context about the root business problem is always needed. Smaller projects may use a single specification document but it must still be capable of being validated by stakeholders and of communicating the requirements to developers in sufficient detail.

To serve its purpose, an SRS must have a number of qualities, some of which derive from the quality of analysis and some of which are construction issues. Its primary job is to specify the system or software completely and unambiguously. The developer should not have to infer the requirements engineer's intentions or the stakeholders' needs.

In most specifications, requirements are written in natural languages such as English to make them understandable by developers and stakeholders. However, care is needed to write requirements in ways that are precise, concise and with only one sensible interpretation. It is good practice to write requirements as single sentences, with the minimum number of conjunctions and using modal verbs consistently. Sometimes, restricted subsets of e.g., English, are imposed to restrict the vocabulary. The following requirement illustrates several of these characteristics:

> *Arm011 — On completion of the arming sequence, there shall be a time delay equal to the escape period before the alarm enters the armed state.*

Several features are illustrated by this example:

- The requirement's interpretation requires the terms *arming sequence, escape period* and *armed state* to be defined in a glossary.

- It is good practice to adopt a standard convention for using modal verbs. For example, the modal verb *shall* is normally used to denote an essential requirement.

- *Arm011* is the requirement's unique identifier (*see Requirements Analysis, Paragraph 5*). The mnemonics chosen for requirement IDs often reflect their derivation.

Despite the expressiveness of natural language, some features required of the system may be difficult to describe precisely; formal notations, such as Z and CSP (mentioned above in Requirements Analysis) provide precise descriptions (in addition to enabling reasoning and proof) because they have formal semantics. In general, however, the software industry has remained skeptical of the benefits of formal notations.

Despite this, there are pockets of success, particularly in critical domains where the perceived additional costs can be offset against the real benefits that the additional rigor of formal specification can bring. Perhaps the most successful notations are those developed to address the needs of an application domain and where training, tools and standardization efforts have acted to further cement the notations' success. A well-known example of this is SDL [Mitschele-Thiel 2001], which is widely used in telecommunications domains.

In general, the requirements engineer should seek the most precise form of description that is consistent with readability and the overall coherence of the document (an SRS that used many different specification languages would be incoherent). This may mean supplementing textual requirements with other formal or graphical descriptions. Most SRSs, for example, include a number of key system models to supplement the textual requirements and help interpret them. Many SRSs allocate a separate section for non-functional requirements.

Requirements analysis must select suitable metrics for non-functional requirements. There is no point, for example, in specifying that a burglar alarm must be "reliable" or "very reliable." These vague, unquantified terms are of little help to the developer and make it impossible to verify the implementation against the specification. Instead, a suitable metric has to be selected such as the alarm's *probability of failure on demand.*

In many cases, functional requirements are derived from a non-functional requirement. For example, a quantified reliability requirement may appear in a system specification while the corresponding SRS may instead specify requirements for a redundant architecture that will satisfy the reliability requirement.

As the formal record of the system specification, an SRS must be a structured document. It must record not only the requirements and associated system models, but should also include (as a minimum) contextual and supporting information, descriptions of constraints, conventions and assumptions, and definitions of specialist terms. A development organization should adopt or adapt a standard SRS format for use in all development projects to help continuity of experience and quality.

7. Requirements Validation

Requirements validation can be crudely characterized as ensuring correctness. The set of requirements specified in the requirements specification must accurately reflect what is needed to solve the underlying business problem, subject to the various constraints that act upon them. This concerns not just the correctness of individual requirements, but the correctness, completenes and consistency of the specification as a whole.

The requirements must also conform to appropriate standards, guidelines and conventions in order to ensure the readability, maintainability, consistency, and other important qualities of a specification document. Strictly speaking, this is about verification rather than validation since it concerns the requirements' conformance to a specification in the form of the standards guidelines, and so forth. Nevertheless, it is important since, for example, poorly documented requirements are hard to validate. IEEE Standard 830-1998 provides a definitive guide to requirement qualities.

Validation is always applied to the requirements specification document, or a final draft of it. However, a more informal kind of validation of the correctness of the requirements is typically performed during elicitation and analysis. Clearly, it is silly to defer all validation until late in the process.

Once validated and any necessary changes are made, the requirements specification is "signed off." Since the effect of this is to issue the document as the basis for design and implementation, both the document and the requirements it contains must be subject to formal change and version control from this point (*see Paragraps 8.1 and 8.2*).

Not all projects are as cleanly organized as this implies, and sometimes development begins before the specification is signed off (a formal signing off may not occur either). For example, an in-house project may be free of the contractual commitment implicit in a signing-off. Similarly, the implementation of certain requirements may be safely anticipated without waiting for a formal commitment from the project sponsor.

However, it is highly desirable to have a formal point in the process at which the requirements are frozen and agreement sought from all the stakeholders. It is essential that change and version control be applied. The absence of a formal signing-off considerably complicates these management tasks.

In most cases, the requirements are validated statically. In some cases where complex dynamic behavior is specified, the requirements may need to be validated dynamically using prototypes or simulations. These are usually costly, however, and a well-run project will anticipate the need to perform this kind of validation well before issue of the draft specification document.

Requirements reviews are a mechanism for validating requirements that should be applied to all requirements. These may take the form of preparatory work in which the reviewers have access to the specification document, followed by a plenary meeting in which the reviewers and the requirements engineers examine key aspects of the requirements. The composition of the review panel is important and it should include both development and stakeholder representatives.

The tasks of review panels can be assisted by including checklists of things to look for *(see [Sommerville & Sawyer 1997] for example)*. This can be particularly useful for checking the requirements specification's compliance with standards or conventions and for generic quality problems, such as weak or ambiguous requirements. A number of tools exist that can help check compliance of the SRS for quality guidelines. For example, NASA's ARM tool measures an SRS against NASA standards using a set of quality metrics [Rosenburg, Hammer & Huffman 1998].

At some stage, the developed system will have to be verified for compliance with the requirements. This is not strictly an issue of requirements validation. However, writing test cases against the requirements is a good way to validate their correctness and clarity. If it proves excessively hard to plan how a requirement is to be verified, then there is likely to be something wrong with the requirement. Although some types of non-functional requirements, such as those concerning security or reliability, are inherently hard to verify, they must be verifiable if they are to serve any useful purpose.

8. Requirements Management

Wiegers [2003] defines *requirements management* as comprising four tasks: change control, version control, requirements tracing, and status tracking. A fundamental prerequisite for each of these is that requirements are uniquely identified (*see Paragraph 4*).

Requirements management is a crucial but often neglected task. Fortunately, significant improvements in industry requirements management practice were stimulated by the issue of the Capability Maturity Model for software (SW-CMM [Paulk et al. 1993]) in the early 1990s. The SW-CMM identified requirements management as a level 2 *Key Process Area.* This effectively meant that all development organizations that wished to demonstrate minimally effective project control and management had to adopt formal requirements management practices. This in turn stimulated a market for requirements management tools.

8.1 Change control.

As noted earlier in this article, *change* (e.g., changes to existing requirements, the emergence of new requirements, changes to requirements' priorities) will occur, even after the SRS has been signed off [Hutchings & Knox 1995]. However, it is crucial that change not be permitted to occur without control. *Requirements creep* is a well-known phenomenon of software projects. Here, uncontrolled or ad-hoc changes to the requirements make project plans impossible to manage, inevitably resulting in systems that are late and over budget. If they ever reach service, they prove to be thoroughly unsatisfactory.

A request to change a requirement needs to be carefully assessed to determine the positive and negative effects of the proposed change. The outcome of this assessment may be a decision to accept or reject the change or perhaps defer it until a subsequent release of the software. *Change control* requires that a formal process be in place to ensure that all the necessary information is available in order to make an informed assessment of the costs and benefits, and that the stakeholders who will form the panel to assess the change request are identified and qualified.

The factors that have to be considered by the assessment panel include how the change will affect project schedule and costs. For example, the effects of a change to a system requirement will propagate through all levels of derived requirements and into the design and implementation components charged with implementing them. If implementation is already underway or (worse) completed, the change will be very costly. Of course, the cost implications of *not* approving the change must also be considered.

Approved changes must be prioritized and the development schedule adjusted to accommodate them. This may necessitate a system delay to the delivery date or failure to implement other, less critical requirements. Negotiation with key stakeholders and reference to the priority and rationale (see *Requirements Analysis, Paragraph 5*) attributes of the requirements affected by a proposed change are crucial.

8.2 Version control.

Requirements changes should be recorded. This should include the fact of the change, the date of approval and the rationale for the change, including the rationale for the decision to approve the change (i.e., *version control*). These notes should form attributes of all affected requirements so that an explicit record of the evolution of the requirements can be maintained for contractual and project post-mortem purposes.

Changes need to be communicated to the development team and this requires the issuing of new versions of the requirements specification. A numbering system for document versions is essential. Several requirements management tools integrate a requirements database with fields for recording requirements and their attributes with document management tools.

8.3 Requirements tracing.

To enable change control and status *tracking*, requirements must be traced. This means that the derivation relationships between requirements should be recorded. Conceptually, the traced requirements form an acyclic graph. Figure 5 shows the derivation relationships between requirements and the allocation of requirements to architectural components.

Requirements must be *traceable* in either direction. *Forward tracing* is necessary to assess the impact of a requirement down through its derived requirements and into the components to which they are allocated. In Figure 5, for example, a change to the system requirement at the bottom of the diagram propagates into two architectural components.

The impact of this change needs to be assessed for each of the elided derived requirements and the affected components. *Backward tracing* is needed for cases where, for example, a change is proposed (perhaps for pragmatic reasons) in a derived requirement or component. In this case, the impact has to be assessed on the system requirements, their stakeholders and ultimately on the system's ability to adequately solve the business problem.

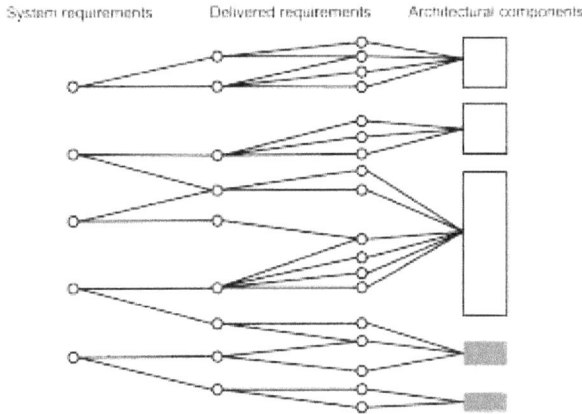

Figure 5: Requirements tracing

References

- **[Beck 2000]** K. Beck, *Extreme Programming Explained: Embrace Change,* Addison-Wesley, 2000.

- **[Finkelsein & Gotel 1995]** A. Finkelstein and O. Gotel, "Contribution Structures," *Proceedings, 2nd IEEE International Symposium on Requirements Engineering (RE '95),* York, UK, 1995.

- **[Hutchings & Knox 1995]** A. Hutchings, and S. Knox, "Creating Products Customers Demand," *Communications of the ACM*, 38 (5), 1995.

- **[IEEE 1998a]** IEEE Standard 1362-1998, "IEEE Guide for Information Technology—System Definition—Concept of Operations (ConOps)" Document, IEEE Computer Society Press, 1998.

- **[IEEE 1098b]** IEEE Standard 1233-1998, "IEEE Guide for Developing System Requirements Specifications," *IEEE Computer Society Press,* 1998.

- **[IEEE 1998c]** IEEE Standard 830-1998, "IEEE Recommended Practice for Software Requirements," *IEEE Computer Society Press*, 1998.

- **[IEEE 1362]** [IEEE Standard 1362-1998].

- **[Karlsson & Ryan 1997]** J. Karlsson and K. Ryan, "A Cost-Value Approach for Prioritizing Requirements," *IEEE Software*, 14 (5), September 1997.

- **[Mitschele-Thiel 2001]** A. Mitschele-Thiel, *Systems Engineering with SDL: Developing Performance-Critical Communications Systems,* John Wiley, 2001.

- **[Paulk et al. 1993]** M. Paulk, W. Curtis, M.B. Chrissis and C. Weber, *Capability Maturity Model, Version 1.1,* IEEE Computer Society Press, 1993.

- **[Roberston & Roberston 1999]** S. Roberston and J. Robertson, M*astering the Requirements Process,* Addison-Wesley, 1999.

- **[Rosenburg, Hammer & Huffman 1998]** L. Rosenburg, T. Hammer and L. Huffman, "Requirements, Testing & Metrics," *Proceedings, 15th Annual Pacific Northwest Software Quality Conference,* Utah, USA, 1998.

- **[Rumbaugh, Jacobson & Booch 1999]** J. Rumbaugh, I. Jacobson and G. Booch, *The Unified Modeling Language Reference Manual,* Addison-Wesley, 1999.

- **[Schneider 1999]** S. Schneider, *Concurrent and Real-time Systems: The CSP Approach,* John Wiley, 1999.

- **[Sommerville & Sawyer 1997]** I. Sommerville and P. Sawyer, *Requirements Engineering—A Good Practice Guide,* John Wiley, 1997.

- **[SWEBOK 2004]** *Guide to the SWE Body of Knowledge,* IEEE, New York, 2004.

- **[Wiegers 2003]** K. Wiegers, *Software Requirements,* Microsoft Press, 2003.

- **[Woodcock & Davies 1996]** J. Woodcock and J.C.P. Davies, *Using Z: Specification, Refinement, and Proof,* Prentice-Hall, 1996.

- **[Young 2001]** R. Young, *Effective Requirements Practices,* Addison-Wesley, 2001.

- **[Yourdon 1989]** Edward Yourdon, *Modern Structured Analysis,* Prentice-Hall,1989.

Chapter 3
Tracing and Traceability

James D. Palmer
George Mason University

Abstract

Traceability gives essential assistance in understanding the relationships that exist within and across software requirements, design, and implementation and is critical to the development process by providing a means of ascertaining how and why system development products satisfy stakeholder requirements, especially for large complex systems. Traceability provides a path to the validation and verification of stakeholder requirements to assure these needs are met by the delivered system, as well as information on testing procedures, performance measures, non-functional characteristics, and behavioral aspects for the delivered system. Both syntactic and semantic information are needed to successfully implement tracing. It is not enough to know the form; it is also necessary to know the substance of the entities to be traced.

However, traceability is often misunderstood, frequently misapplied, and seldom performed correctly. There are many challenges to achieving traceability, particularly the absence of automated techniques to assist in the identification of linkages from requirements to design, or test, or operation needed to trace entities within and across the system development process. A particular challenge to providing traceability to and from system level requirements is that it becomes necessary to utilize both the constructs of language semantics as well as syntax.

Traceability is introduced, and its place in a development process, coupled with the values and pitfalls, are covered. The essentials of traceability are examined together with how to implement tracing within a development life cycle for large complex systems. Working definitions and related terms are provided to assure common understanding of the terminology and application of tracing in system and software development. A review of contemporary approaches to implement tracing with an overview of several of the Computer Supported Software (or System) Engineering (CASE) tools that purport to support tracing are given and future trends are examined.

1. Introduction

Successful system development depends on the ability to satisfy stakeholder needs and requirements and to reflect these in the delivered system. Requirements, design and implementation that are complete, correct, consistent, and error free, play a major role in ensuring that the delivered system meets stakeholder needs. Critical keys to this are understanding and tracing the relationships that exist amongst system requirements, design, code, test, and implementation. Large-scale complex systems are initiated by stakeholder determination that a need exists that is not met by existing systems. From this beginning, system level requirements are developed to broadly outline the desired capabilities, which, in turn, are investigated to ascertain feasibility and practicality and to examine trade-offs. Once the feasibility and practicality of the desired system have been determined to be necessary and sufficient to launch a new system (or significant modification of an existing or legacy system), design is completed, and systems are constructed, tested and fielded. It is essential to maintain traceability from the system requirements to operation and maintenance to assure that the delivered system meets the stated organizational needs of the stakeholder.

2. System Life Cycle for Traceability Management

Generally, a system or process development life cycle is followed to produce the desired system. There are many life cycle models [1], and one of the simplest is the system development or waterfall life cycle model depicted in Figure 1. It also serves as the basis for most life cycle models in use today, such as the spiral model, the evolutionary model, and the prototyping model. Within any system development life cycle, requirements must be traced both forward and backward to assure that the correct system is being designed and produced, and that the correct design and production approaches are used.

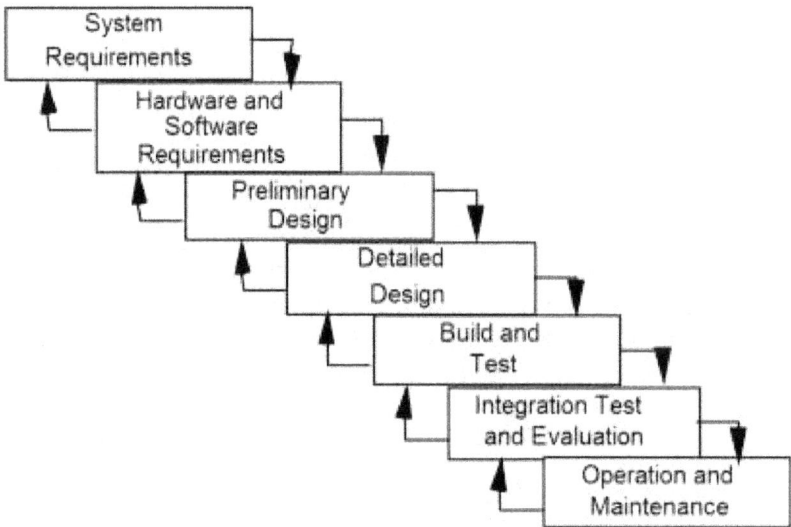

Figure 1: Typical system and software development life cycle

In the life cycle model of Figure 1, system requirements, usually prepared by James D. Palmer in natural language, are provided by the stakeholder to the developer. These system requirements, if they exist at all, may be poorly written and only vaguely define stakeholder desires for the new system. This may impact the ability to construct a system that will satisfy the stakeholder. From these system requirements, hardware and software requirements and specifications are prepared. Requirements and specification development are followed by preliminary design; detailed design; construction of the system including hardware and software; system integration, testing and evaluation; and finally, installation including operation and maintenance.

These life cycle activities require documentation of needs and outcomes. Each must trace forward to the subsequent activity and backward to the preceding one. Clearly, traceability, both forward and backward, is essential to verify that the requirements of one phase translate to outcomes of that phase which become the requirements for the next phase, and so on through the development activity. Traceability is equally essential to validate that system requirements are satisfied during operation.

3. Need for Traceability

Traceability is essential to verification and validation and is needed to better understand the processes used to develop the system and the products that result. It is needed for quick access to information, information abstraction, and to provide visualization into the techniques used for system development. Traceability is needed for change control, development process control and risk control.

Tracing provides insights to non-behavioral components such as quality, consistency, completeness, impact analysis, system evolution, and process improvement. It is equally important to have the capability to trace a requirement or design or code module to its origin and to test it. Stakeholders recognize the value of properly tracing within and across the entities of a system through risk management insights, appropriate integration tests, and the delivery of a project that meets the statements of the requirements [2].

Traceability supports assessment of under- or over designs and investigation of high-level behavior impact on detailed specifications, as well as non-functional requirements such as performance and quality factors. Moreover, traceability supports conflict detection by making it feasible to examine linkages within and across selected entities and by providing visibility into the entire system. Through tracing, there is the assurance that decisions made later in the system development life cycle are consistent with earlier decisions. Test cases check that coverage for code and integration testing and for requirements validation is provided. Traceability provides the basis for the development of an audit trail for the entire project by establishing the links within and across system entities, functions, behavior, and performance.

While there is widespread acceptance of the necessity to trace, there is considerable controversy as to the ultimate need, purpose and cost of tracing from requirements to the delivered product. The controversy arises primarily because of the lack of automated approaches to implement the process and the concomitant time and effort that must be applied with any of the presently available support tools. Developers simply do not see the benefits that may accrue to the final product when traceability is fully implemented compared to the time and effort required.

4. Problems and Issues Concerning Traceability

Difficulties related to tracing generally revolve around the necessity to manually add trace elements to requirements documents and subsequent work products from software development. Since these products have little or no direct consequence to the development team, assignment of trace elements generally has a low priority. The benefits of traceability are not seen until much later in the development life cycle, usually during validation testing and system installation and operation, and then primarily by integration testers and stakeholders rather than developers. Additionally, traceability is often misunderstood, frequently misapplied and seldom performed correctly.

Issues and concerns emanate from the complexity of a project itself that must be confronted when implementing traceability. Each discipline, such as avionics, communications, navigation, security, or safety may have languages, methods and tools peculiar to the discipline. This results in a lack of ability to trace across disciplines, which, in turn, may lead to errors in traceability matrices used to provide linkages within and across disciplines.

Some of the issues that need to be addressed by the stakeholder and developer at the time of system development include how to apportion projects by discipline, the type and nature of information that should be traced across different disciplines, and the types of tools that can be used to provide consistent and correct traceability across disciplines. Establishing threads across disciplines is also difficult due to language, method and tool peculiarities.

Currently, there is no single modeling method or language sufficiently rich to represent all aspects of a large complex system and still be understandable to those involved. In tracing information across different disciplines and toolsets, and to provide threads across these, essential system properties and the classification schemes used are needed. Such properties and schemas do not usually exist. Thus, for verification and validation, traceability must always focus on a common denominator, that of the approved system requirements. Finally, internal consistency of the baseline documentation may not be adequate to support tracing. This latter properity is usually a significant problem in the modification of legacy systems.

4.1 Definitions of terms.

There are many terms that describe, delineate or relate to traceability. Some of these correlate to the "how and why" for traceability, while others connect to the outcomes or "what" of traceability. In general, the basic meaning of the term is that which is provided by Webster's New Collegiate Dictionary [3], while the last meaning is given in the context of systems and software engineering, as an example of usage.

- **Allocation** — The act of distributing; allotment or apportionment; as to assign or apportion functions to specific modules.

- **Audit** — A formal checking of records, to determine that what was stated was accomplished; to examine and verify; as to confirm a stated capability is met in the software product.

- **Behavior** — The way in which a system acts, especially in response to a stimulus; stimulus-response mechanisms; as activity or change in reliability across sub-systems.

- **Bottom-up** — A design philosophy or policy that dictates the form and partitioning of the system from the basic functions that the system is to perform and moving up to the top-level requirements; as a design policy that provides basic modules followed by top-level constructs.

- **Classification** — A group of entities ranked together as possessing common characteristics or quality; the act of grouping or segregating into classes which have systematic relationships; a systematic grouping of entities based upon some definite scheme; to classify requirements according to organizational or performance characteristics.

- **Flow down** — To move or circulate from upper to lower levels; as to trace a requirement from a top-level to designs to code to test.

- **Function** — The characteristic action or the normal or special action of a system; one aspect of a system is so related to another that there is a correspondence from one to the other when an action is taken; as an algorithm to provide the equations of motion.

- **Hierarchy** — A series of objects or items divided or classified in ranks or orders; as in a type of structure in which each element or block has a level number (1= highest), and each element is associated with one or more elements at the next higher level and lower levels; as a single high level requirement decomposes to lower level requirements and to design and code.

- **Impact analysis** — Separation into constituent parts to examine or distinguish contact of one to another, a communicating force; as to focus on software changes and the traceable consequences; relating software requirements to design components.

- **Policy** — Management or procedure based primarily on material interest; as a settled course or level to be followed for system security.

- **Requirement** — A requisite condition; a required quality; to demand; to claim as by right or authority; to exact; as to demand system performance by the stakeholder.

- **Thread** — To connect; as to pass a thread through; string together; as to link behaviors of a system together.

- **Top-down** — A design philosophy or policy that dictates the form and partitioning of the system from the top-level requirements perspective to the lower level design components; as in a design policy for all activities from high-level requirements to design and code.

- **Top-level requirement** — A requisite condition leveled by the stakeholder; as a system level requirement for security.

- **Traceability** — Requirements traceability is concerned with documenting the life of a requirement and providing bi-directional traceability between various associated requirements. It enables users to find the origin of each requirement and track every change that was made to this requirement. For this purpose, it may be necessary to document every change made to the requirement.

- **Traceability management** — To control and direct; guide; administer; give direction to accomplish an end; as to control and direct tracing from top level through to design and code.

- **Tracing** — Tracing involves a specialized use of logging to record information about a program's execution.

- **Tree** — A diagrammatic representation that indicates branching from an original stem; as software components derived from a higher level entity to more discrete lower level entities.

4.2 State of the practice of traceability.

Traceability management applies to the entire development life cycle from project initiation through operation and maintenance as shown in Figure 2.

It is presently feasible to manage tracing using a combination of manual and automated assistance, thus providing some assurance that the development of a system meets the needs as provided by the stakeholder. An essential element of successful traceability management, provided by currently available CASE tools, is the ability to provide links from requirements forward to designs, code, test, and implementation, and backward from any of these activities to requirements once these links have been manually entered into the CASE tool.

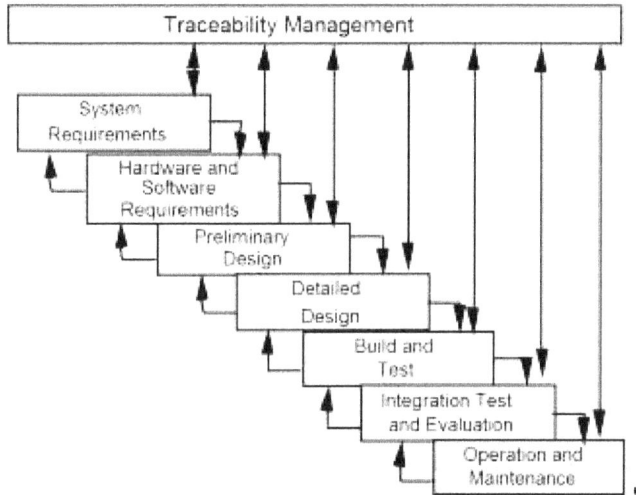

Figure 2: Traceability management across the system development life cycle

Techniques currently in use to establish and maintain traceability from requirements through design, code, test, and operation begin with manual identification of linkages. These linkages may be subsequently supported by document managers, a database or CASE tools specifically designed for requirements traceability management.

5. Contemporary Traceability Practices

Traceability has traditionally been accomplished by manually assigning and linking unique identifiers; that is, a sentence or paragraph (or other partition) requirement is assigned a particular alpha-numeric reference. This information is subsequently managed in a word processor or database, often through use of a CASE tool. Even with the use of a CASE tool, the initial identification of trace entities and linkages must be accomplished manually. By establishing a unique identification system and following this scheme throughout the life of the project, it is possible to trace these specific entities both forward and backward from requirements to product. This unique identity may be linked within and across documents using manually derived traceability tables to assure full traceability over all aspects of the project.

A typical output of tracing is a traceability matrix that links high-level requirements to each and every other requirement or specification of the system. A typical traceability table for a large complex system is shown in Table 1. In this table, individual requirements in the Systems Requirements Document (SRD) have been manually linked to more detailed system requirements in the Systems Specification, which in turn have been manually linked to particular specifications in the system segments.

Table 1: Traceability matrix for a multi-segment system

SRD	SS	Segment	Segment	Segment	ICD
3.1.2.1	3.3.4.5	3.2.2.5.6	3.5.3.2		3.1.4.6.7
	3.3.4.6	3.2.2.5.7			3.1.4.6.8
		3.4.5.6.2			3.1.4.6.9
3.4.3.1	3.6.7.2	3.5.2.5.1	3.7.4.3.1	3.6.4.5.2	3.3.2.4.5
	3.8.4.3		3.7.4.3.2		3.3.2.4.7

Other matrices or tables may provide more details such as cryptic messages, partial text, critical values, or the entire text. The system represented in the traceability table is configured as in Figure 3. The SRD represents stakeholder input, the SS represents the initial interpretation of these high level requirements by developers, and the segment specifications provide more detailed information for the design.

Traceability is especially critical for the operation and maintenance phase. This is when significant stakeholder changes may be made and change impacts and impact analyses must be performed. Such changes are difficult to trace; however, without tracing it is nearly impossible to ascertain the extent of the full impact of additions, deletions or modifications to the system.

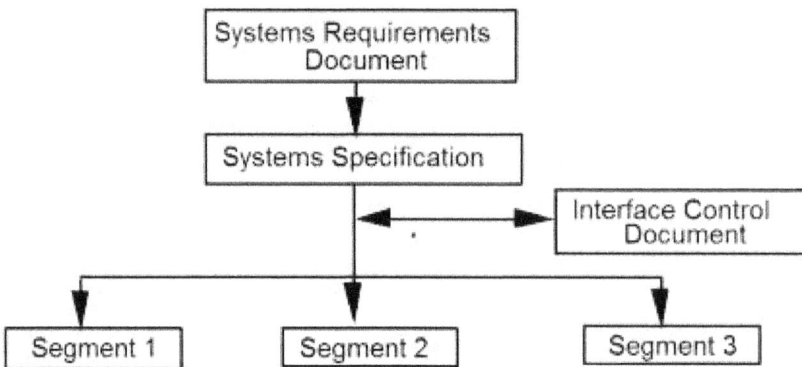

Figure 3: Typical requirements classification schema for a large complex system

6. An Ideal Process for Traceability

To understand what must be traced, we need a defined process for developing system architectural views and classification schemes, as well as processes for specifying and verifying the products to be constructed. This is generally provided by the stakeholder in consort with the developer. The development of these views is necessary to partition the project for design and construction.

An ideal traceability process consists of the steps of identification, architecture selection, classification, allocation, and flow down as depicted in Figure 4.

The process begins with the identification of requirements at the system level, specification of system architecture and selection of classification schema. Following this, allocations are made in accordance with the selected schema. Following allocation, the requirements flow down to design, code and test. This top-down approach has proven most effective in the management of traceability for large-scale complex projects.

However, this approach is basically a manual activity that requires significant investment of time and effort on the part of skilled personnel. The outcomes represent a system hierarchy along the lines of the classification structure used for the architectural allocations. It is also necessary to provide threads through the various behavioral and non-behavioral aspects of the project to complete the traceability process. These thread paths are manually assigned using approaches such as entity-relation-attribute diagrams. For example, tests are threaded back to requirements through code and design.

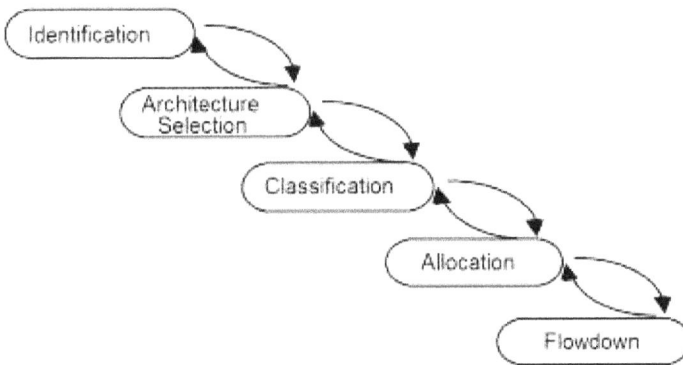

Figure 4: The ideal traceability process

Once the system hierarchy, the architecture and classification schema have been defined, identified system requirements are assigned to the top-level block of the hierarchy. At this time, they are added to the traceability database for storage, retrieval and reuse. After appropriate analyses, these requirements are decomposed and flow down into more detailed requirements for each of the lower level blocks to which the requirement was allocated, as was shown in the example of Figure 3. The higher level requirements are sometimes referred to as parents and the lower level ones as children. Change notification should be rigorously traced to determine the impact of such activities on changes to cost, schedule and feasibility of system design and implementation on tests that must be conducted and on support software and hardware.

7. Actual Practice for Implementing Traceability

In actual practice, tracing is a labor intensive and aggravating task. Domain experts follow a process to decompose the system that is similar to that depicted in Figure 3. Once appropriate systems architectures are identified, a classification schema or schemas for purposes of allocation of requirements to system specific architectures is prepared and requirements are assigned to specific units.

Refering to examples of the types of classification schemas used, one may be centered on functional aspects of the project; such as navigation, communications or threat assessment; another may concentrate on performance and security; while yet another may be focused on stakeholder organization. It is not feasible to enumerate, a priori, all the ways in which the project may need to be partitioned and viewed; thus, traceability becomes a continuous process as perspectives change and as requirements change. To validate these various views, there is only one common basis from which to form trace linkages: the system requirements.

The next step, after receipt of the requirements documents and delineation of the system architecture, is to determine the nature of the tracing to be accomplished. Several options are feasible; these include working with statements that contain "shall," "will," "should," "would," or similar verbs; or with entire paragraphs; or the total set of statements provided by the stakeholder. The strongest selection is "shall" statements, which may be the only contractually acceptable designation for a requirement. This is followed by the development of classification schemas according to function, data object, behavior, organization, or other approaches. Once the option(s) has been selected, the requirements are parsed according to the option and assigned a unique identity. For example, if "shall" has been selected as the option, sentences with "shall" as a verb are collected and are identified sequentially, while also retaining the original identification system provided by the stakeholder. This new identification system is maintained throughout the life of the project.

Syntactic and semantic information are both necessary to perform tracing. Language semantics are needed to assure the trace is related to the meaning or context of the requirement or set of requirements, while syntax is necessary to trace to a specific word or phrase, without regard to meaning or context. Integration of both constructs is required to provide for full traceability from natural language statements to the other steps as shown in Figure 2. Manual verification of outcomes is required to assure compliance with the intent and purpose of the tracing activity.

Next comes allocation according to the classification schema. This likewise is a manual task, even with automated assistance from one of the available CASE tools, as most of these tools require the operator to physically establish the links from one entity to another for traceability. All linkages must be designated and maintained and traceability matrices are generated from these outcomes.

If a CASE tool has been used that supports generation of traceability matrices, these are created automatically; otherwise, these matrices must be manually prepared. These steps are depicted graphically in Figure 5. These results are usually stored in a traceability database. The traceability linkages are subsequently designated and maintained across the entire development project from design to code to test to operation and maintenance.

The benefits are largely intangible and are related to the avoided costs associated with rework and possible failure of the product to satisfy stakeholders. To estimate the benefits, it would be necessary to prepare various scenarios, simulate the outcomes due to failure of various aspects of the development process, and estimate the value of avoiding these failures.

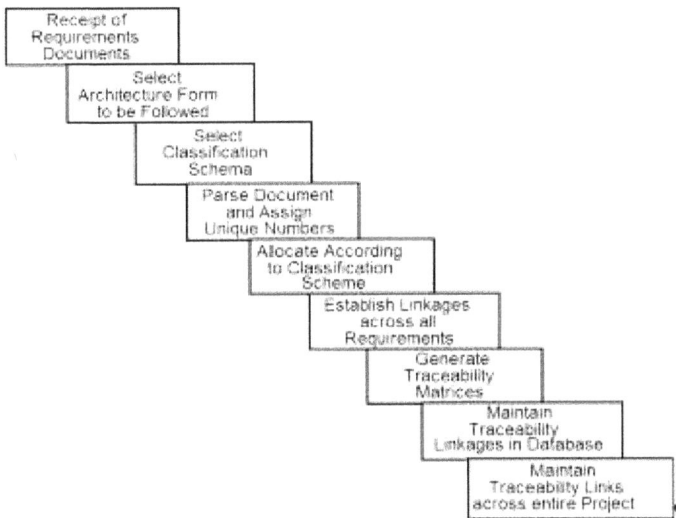

Figure 5: Steps to accomplish traceability

8. Return on Investment for Traceability

Risk factors must also be taken into consideration in calculation of the potential benefits, including the potential that the project will not meet stakeholder needs. Assessing benefits without comparative analyses is generally not possible. Generating this information is considered to be unfeasible due to the costs of running such experiments and the need to develop realistic scenarios that may or may not ever be replicated in actual practice.

9. Current Traceability Tools

Typical of the currently available automated (or semi-automated) assistance approaches to traceability are those that provide for traceability through a variety of syntactic language components: hypertext linking, unique identifiers, syntactical similarity coefficients, or combinations of these. In hypertext linking, the "hotword" or word/phrase to be linked to other requirements is manually identified and entered into the hypertext tool. Links are automatically made and maintained by the tool to provide forward and reverse traceability for the word selection. In the unique identifier approach, an identifier is assigned that remains with the individual requirement throughout the life of the project.

To assure traceability, this unique identifier provides a "fan-out" capability within a hierarchical structure such that one system level ("A" level) requirement may be the parent to many "B" level requirements which, in turn, may be the parents for great numbers of "C" level requirements, as depicted in Table 1. Use of syntactic similarity coefficients ascertains whether or not a pre-defined set of words of a given requirement are found in another requirement. When the degree of similarity is above a pre-defined threshold, the two requirements in question are said to trace.

There are problems with each of these approaches. They do not consider the semantics or context in which the tracing is to occur. Hypertext linking finds the search text without regard to the placement in the text and without regard to the way in which the words are used. Use of a unique identifier provides access only to those requirements so identified with no perspective as to meaning or context. Syntactic similarity coefficient traceability is like hypertext linking in that it is indiscriminate as to the meaning and context of the requirement to be traced.

Commercially available requirements tools utilize straightforward traceability links that must be manually developed to relate requirements to other requirements and to design, code and implementation. Current methods for implementing traceability with these commercial tools generally involve the manual provision of links within and across documents and then automated management of these documents. Traceability links are used to establish the one-to-one, one-to-many, many-to-one, or many-to-many relationships that may exist, as may be seen from Table 1. As noted previously, linkages are not automatically established by tools during the development process, but must be manually generated. From this point, automated assistance is provided by the tool to manage traceability.

At present, there are no standards available to support tools for traceability, which has led to the development and use of a large number of commercial tools, each with differing methods, as well as proprietary tools developed by certain industries because it is considered to be a competitive advantage for most large complex projects. A number of commercially available tools have been developed to support traceability and a number of general CASE tools provide support to traceability management, especially from requirements forward to design, code, test, and operation. One of the common activities for all tools is manual development of architectural perspectives and classification schemas. Another common feature is the need to manually establish the initial linkages within and across all traceable entities. Once the initial linkages have been established, these tools effectively and efficiently manage a traceability database or word processor document.

10. Common Tool Characteristics

There are some common tool characteristics that are deemed to be minimal to provide support for traceability. The tool must be well understood by and be responsive to users and match the characteristics of the development environment used by the developers. Tools must also accept and utilize the data that is provided in the form provided. In addition, the tool must be flexible, capable of operation in an automated assistance mode to support various activities and services, such as active and passive data checking, batch as well as on-line processing, addition, deletion, and modification of a requirement, customization to specific domain applications, dynamic database structure for change management, and a tailorable user interface. Traceability tools will never be fully automated, as human decision making is essential to the establishment of classification schema and system architecture designation. Human interaction and decision making is both desirable and necessary to maximize the interactions between the project stakeholder and developer of the project.

11. Commercial CASE Tools for Traceability

Some commercially available tools have been developed for traceability link information expressed by a single discipline within a single phase, while others have been developed specifically to link requirements to other activities within the development life cycle. Cadre Team Work for Real-Time Structured Analysis (CADRE) is a tool that operates on a single discipline within a single phase. Tools that link information from multiple disciplines and phases include: Requirements Traceability Manager (RTM) (Marconi Corporation) [4], SLATE (TD Technologies) [5], and DOORS (Zycad Corporation) [6].

These tools use an entity-relation-attribute-like schema to capture information on a system database that is either relational or object-oriented, enable formation of queries about traceable entities, and are used for generating reports. RTM uses a relational database structure to capture information and provide management, while DOORS provides an object-oriented database for management of information. SLATE follows a multi-user, client-server, object-oriented approach that provides dynamic representation of the system as it evolves.

Another method used by commercial tool vendors is the hypertext approach. In this approach, keywords or phrases are identified as being indicative of traces. These are linked through hypertext throughout the document or set of documents that comprise the requirements. An example of a tool that uses this approach is Document Director [7].

Some general-purpose analysis tools are also used for tracing. Some of the more robust tool sets include: Requirements Driven Design (RDD-100 by Ascent Logic) [8], which is used to document system conceptual models and Foresight [9], which is utilized to maintain a data dictionary and to document system simulation.

Other tools and techniques that support requirements traceability include Software Requirements Methodology (SERM); Problem Statement Language/Problem Statement Analyzer (PSL/PSA); N^2 charts; Requirement Networks (R-Nets); and ARTS (a database management system for requirements). Not all of the CASE tools support requirements traceability; however, most do support some form of requirements management.

12. Future Trends and Conclusions

The future in traceability support lies in the development of the capability to deal directly with requirements in natural language, and the ability to provide automated assistance to allocation of requirements to various architectural and classification systems. From this automated assistance it becomes feasible to provide for and manage a traceable baseline for the entire system.

The following issues are being addressed in ongoing research programs:

- Automated allocation of entities to architectures and classifications.

- Traceability that is independent of methods used to develop architectures and classifications.

- Tracing product attributes from requirements to the lowest levels.

Several research programs are working on the problems associated with natural language: the two addressing traceability are from George Mason University and Trident Systems. The Center for Software Systems Engineering at George Mason University has developed and applied an automated assistance approach to the problems of allocation of entities to architectures and classification called the Automated Integrated Requirements Engineering System (AIRES) [10]. Trident Systems intends to develop a CASE tool called RECAP (Requirements Capture) which is intended to manage natural language requirements [11].

AIRES provides an assessment framework and techniques for integrated application of both semantic and syntactic rules for effective, efficient and comprehensive identification of traceable and non-traceable requirements in large, complex multiple-segment systems. The framework provides for the categorization of requirements in classification structures through the application of a diverse combination of rules and procedures, each of which applies unique combinations of both semantic and syntactic classification rules and tables for the categorization of requirements. These serve as the basic building blocks of the assessment framework and may be applied either singly or in combination.

AIRES supports automated development of linkages that may be transferred electronically to commercially available traceability tools such as RTM for management of a requirements database and report generation. AIRES is presently available in prototype form and has been utilized in support of several large complex system developments for traceability support [12].

RECAP, presently a conceptual design, is intended to provide a set of interfaces that permit the operator to manipulate natural language requirements. RECAP proposes to combine the information management and extraction capabilities of information retrieval system approaches with knowledge-base rules. It also intends to provide sequential and string search access to any portion of the document set. Quick access to information is proposed through keywords, sentence identifiers or rule-based queries. The user will be required to provide information for resolution of ambiguity, mistakes in statements and addition of missing items. RECAP is intended to aid the user in making these decisions [11].

Information linked by these tracing tools is not dependent upon a model or discipline. It is possible to link entities as needed; for example, it may be desirable to link the estimated footprint, weight and power usage of a piece of computer equipment (stored in a hardware modeling tool) to the estimated throughput and memory requirements for a piece of software (stored in a software modeling tool). To efficiently use these tracing tools, it is necessary to automatically transfer the information captured to CASE tools used downstream in the development life cycle. This is accomplished by tracing system definitions, system development processes and interrelationships across system units.

While tracing from origination to final product is a difficult and arduous, manually intensive task at the present time, advances in technology should soon be commercially available to assist in automated allocation and classification procedures. These advances will make the traceability task much more reasonable, feasible and supportable for large complex system developments due to the automated assistance provided for allocation and classification, the most labor intensive aspects of tracing. In each of the approaches, the CASE tool provides automated assistance to tracing, but requires human operator inputs only for decision-making activities. These tools represent a significant advance over the present state of the practice for traceability.

References

[1] Sage, Andrew P. and Palmer, James D., *Software Systems Engineering*, John Wiley and Sons, 1990.

[2] White, Stephanie, "Tracing Product and Process Information when Developing Complex Systems," *CSESAW '94* July 19-20, 1994, pp. 45-50, NSWCDD/MP-94/122.

[3] *Webster's New Collegiate Dictionary*, Sixth Edition, G. & C. Merriam Co., Springfield, MA, 1951.

[4] "RTM-Requirements & Traceability Management, Practical Workbook," GEC-Marconi Limited, October, 1993.

[5] Nallon, John, "Implementation of NSWC Requirements Traceability Models," *CSESAW*, White Oak, MD, July 19-20, 1994, pp. 15-22, NSWCDD/MP-94/122.

[6] Rundley, Nancy and Miller, William D., "DOORS to the Digitized Battlefield: Managing Requirements Discovery and Traceability," *CSESAW*, White Oak, MD, July 19-20, 1994, pp. 23-28. White, Stephanie, "Tracing Product and Process Information when Developing Complex Systems," *CSESAW*, White Oak, MD, July 19-20, 1994, pp. 45-50, NSWCDD/MP-94/122.

[7] "Document Director-The Requirements Tool," B.G. Jackson Associates, 17629 E. Camino Real, Suite 720, Houston, TX 77058, 1989.

[8] "RDD-100-Release Notes Release 3.0.2.1, October, 1992", Requirements Driven Design, Ascent Logic Corporation, 180 Rose Orchard Way, #200, San Jose, CA 95134, 1992.

[9] Vertal, Michael D., "Extending IDEF: Improving Complex Systems with Executable Modeling," *1994 Annual Conference for Business Re-engineering*, IDEF Users Group, Richmond, VA, May, 1994.

[10] Palmer, James D. and Evans, Richard P., "An Integrated Semantic and Syntactic Framework for Requirements Traceability: Experience with System Level Requirements for a Large Complex Multi-Segment Project," *CSESAW '94* July 19-20, 1994, pp. 9-14, NSWCDD/MP-94/122.

[11] Hugue, Michelle, Casey, Michael, Wood, Glenn, and Edwards, Edward, "RECAP: A Requirements Capture Tool for Large Complex Systems", *CSESAW*, White Oak, MD, July 19-20, 1994, pp. 39-44, NSWCDD/MP-94/122.

[12] Palmer, James D. and Evans, Richard P., "Software Risk Management: Requirements-Based Risk Metrics," *Proceedings of the IEEE 1994 International Conference on SMC*, Oct. 2-6, 1994, San Antonio, TX.

Chapter 4
Software Prototyping

Wikipedia, The Free Encyclopedia

Abstract

Software prototyping is the activity of creating prototypes of software applications during software development. Prototypes are incomplete versions of the software program being developed software. Prototyping is comparable to prototyping in other fields, such as mechanical engineering or manufacturing.

A prototype typically simulates only a few aspects of the project being developed and may be completely different from the final product.

Prototyping has several benefits: The software designer and implementer can obtain valuable feedback from users in the early project phases. The client and the contractor can compare whether or not the software product matches the determined software specifications according to which the software program is built. Prototyping also allows the software engineer to determine the accuracy of initial project estimates and whether proposed deadlines and milestones can be successfully met. The degree of completeness and the techniques used in prototyping have been in development and debate since its proposal in the early 1970s.

1. Overview

The original purpose of a prototype is to allow users of the software to evaluate the developer's design description of the eventual product. This evaluation is conducted by testing the proposals, rather than interpreting and evaluating the design based on descriptions. Prototyping can also be used by end users to describe and prove requirements that have not been considered and that can be key factors in the commercial relationship between developers and their client.

This prototype process is in contrast to the 1960s and 1970s monolithic development cycle of first building the entire software program and then working out any inconsistencies between design and implementation, which led to poor estimates of time and cost and resulting higher software costs. The monolithic approach has been dubbed the "Slaying the (Software) Dragon" technique, because it assumes that the software designer and developer together comprise a single hero who has to slay the entire dragon alone. Prototyping can also be used to avoid the great expense and difficulty of changing a finished software product.

The practice of prototyping is one of the points Frederick P. Brooks discusses in his 1975 book, *The Mythical Man-Month* [1975], and his 10-year anniversary edition.

The process of prototyping involves the following four steps:

1. ***Identify basic requirements*** — Determine basic requirements including the input and output information desired. Details, such as security, can typically be ignored.

2. ***Develop initial prototype*** — The initial developed prototype includes only user interfaces. (*See Horizontal prototype below.*)

3. ***Review*** — The customers, including end-users, examine the prototype and provide feedback concerning proposed additions or changes.

4. ***Revise and enhance the prototype*** — Make changes and improvements as necessary.

Both the initial specifications and the prototype can be improved. It may be necessary to negotiate what is contained within the scope of the contract or product. If changes are introduced, then a repeat of steps #3 and #4 may be needed.

2. Types of Prototypes

Jacob Nielsen summarizes the various dimensions of prototypes in his book, *Usability Engineering* [1993].

2.1 Horizontal prototype.

A common term for a user interface prototype is the *horizontal prototype*. It provides a broad view of an entire system or subsystem—focusing on user interaction more than low-level system functionality, such as database access. Horizontal prototypes are useful for:

- Confirming user interface requirements and system scope.

- Demonstrating a version of the system to obtain buy-in from the customer or user.

- Developing preliminary estimates of development time, cost and effort.

2.2 Vertical prototype.

A *vertical prototype* is a more complete elaboration of a single subsystem or function. It is useful for obtaining detailed requirements relating to a given function with the following benefits:

- Refined database design.

- Obtain information on data volumes and system interface needs for network sizing and performance engineering.

- Clarify complex requirements by "drilling down" to actual system functionality.

2.3 Other types of prototyping.

Software prototyping has many variants. However, all prototype methods are based on two major types of prototyping: *throwaway prototyping* and *evolutionary prototyping.*

2.3.1 Throwaway prototyping.
Also called *rapid* prototyping and *close-ended* prototyping, *throwaway* prototyping refers to the creation of a model that will eventually be discarded rather than becoming part of the final delivered software system. After preliminary requirements gathering is accomplished, a simple working model of the system is constructed and visually shown to the users, demonstrating what the requirements may look like when implemented into a finished system.

Throwaway prototyping involves creating a working model of various parts of the system at a very early stage, after a relatively short investigation. The method used in building the prototype is usually quite informal, the most important factor being the speed with which the model is provided. The model then becomes the starting point from which users can re-examine their expectations and clarify their requirements. Once this has been achieved, the prototype model is "thrown away", and the system is formally developed based on the identified requirements [Crinnion 1991].

The most obvious reason for modeling throwaway prototyping is that it can be done quickly. If the users can obtain instant feedback on the stipulated requirements, they may be able to refine them early in the development of the software.

Making changes early in the development life cycle is extremely cost effective since there is little at that stage to be redesigned. If a project is changed after a considerable amount of design work has been done, then small changes could require large efforts to implement since software systems have many dependencies. Speed is crucial in implementing a throwaway prototype, because a limited budget of time and money allows for little to be expended on a prototype that will be discarded.

Another strength of throwaway prototyping is its ability to construct interfaces to be tested by the users. The user sees the user interface as the system. It is much easier to grasp how the system will work by personally viewing it.

It is asserted that evolutionary rapid prototyping is a more effective manner with which to address user requirements-related issues, and therefore, a greater enhancement to software productivity overall. Requirements can be identified, simulated and tested far more quickly and cost effectively when issues of evaluability, maintainability and software structure are not simultaneously addressed. This in turn leads to the accurate specification of requirements and the subsequent construction of a valid and usable system from the user's perspective via conventional software development models.

Prototypes can be classified according to the fidelity with which they resemble the actual product in terms of appearance, interaction and timing. One method of creating a low-fidelity throwaway prototype is *paper prototyping*. The prototype is implemented using paper and pencil, and although it mimics the function of the actual product, it does not look at all like it. Another method of easily building high-fidelity throwaway prototypes is to use a GUI Builder and create a *click dummy*, a prototype that looks like the goal system, but does not provide functionality. The usage of storyboards, animatic (*an animated storyboards*).or drawings are not exactly the same as throwaway prototyping, but are certainly in the same family. These are non-functional implementations that demonstrate the proposed appearance of the propsed system as it is envisioned to be delivered [SPC 1997].

2.3.2 Evolutionary prototyping, *Evolutionary prototypes* have an advantage over throwaway prototypes in that they are functional systems. Although they may not contain all features planned by the users, they may be used on an interim basis until the final system is delivered.

It is not unusual within a prototyping environment for the user to place an initial prototype to practical use while waiting for a more developed version to be designed. The user may decide that a 'flawed' system is better than no system at all [Crinnion 1997].

In evolutionary prototyping, developers can focus on developing parts of the system that they understand instead of developing a whole system.

To minimize risk, the developer does not implement poorly understood features. Once complete, the partial system is sent to customer sites. As users work with the system, they detect opportunities for new features and provide requests for these features to developers. Developers then consider these enhancement requests along with their own suggestions and use sound configuration-management practices to change the software requirements specification, update the design, recode, and retest [Davis 1991].

(1) *Incremental prototyping* — The final product is built as separate prototypes. The separate prototypes are then merged into an overall design. With the help of incremental prototyping,the development time gap can be reduced.

(2) *Extreme prototyping* — Extreme prototyping as a development process is used in particular for developing web applications. Extreme prototyping partitions web development into three phases, each phase based on the preceding one. The first phase is a static prototype that consists mainly of HTML pages. In the second phase, the screens are programmed and become fully functional by using a simulated services layer.

Services are implemented in the third phase. The process is called extreme prototyping to draw attention to the second phase of the process, whereby a fully functional UI is developed with very little regard to the services other than their agreed upon contract.

3. Advantages of prototyping.

There are many advantages to using prototyping in software development– some tangible, some abstract.

(1) ***Reduced time and costs*** — Prototyping can improve the quality of requirements and specifications provided to developers. Because changes cost exponentially more to implement when detected later in the development phase, the early determination of *what the user really wants* can result in software designed in less time and at reduced cost.

(2) ***Improved and increased user involvement*** — Prototyping requires user involvement. This requirement allows users to interact with a prototype, allowing visual interaction to provide better and more comprehensive feedback and specifications.

The fact that the prototype is examined by the user prevents many misunderstandings, and limits miscommunication that occurs when each side erroneously believes the other understands specified stipulations. Since users know the problem domain better than anyone on the development team, increased interaction can result in a final product that contains greater tangible and intangible qualities.

4. Disadvantages of prototyping.

Using, or perhaps misusing, prototyping can also have a number of disadvantages.

(1) ***Insufficient analysis*** — Focus on a limited prototype can distract developers from properly analyzing the complete project. This can lead to overlooking better solutions, preparation of incomplete specifications or the conversion of limited prototypes into poorly engineered final projects that are difficult to maintain. Furthermore, since a prototype is limited in functionality, it may not bode well if the prototype is used as the basis of a final deliverable, an oversight that may not be noticed if developers are overly focused on building a prototype as a model.

(2) ***User confusion of the prototype versus the finished system*** — Users may think that a prototype, intended to be thrown away, is actually a final system that merely needs to be finished or polished. (Users are, for example, often unaware of the effort needed to add error-checking and security features, which a prototype may not contain.) This unawareness can lead users to expect the prototype to accurately model the performance of the final system when this is not the intent of the developers.

Users can also become attached to features that were included in a prototype for consideration and then removed from the specification during implementation of the final system. If users are able to require that all proposed features be included in the final system, the system will be incomplete, error prone, and will lack an underlying architecture and a solid architectural base.

(3) ***Developer misunderstanding of user objectives*** — Developers may assume that users share their objectives (e.g., to deliver core functionality on time and within budget), without understanding wider commercial issues. For example, users attending to enterprise software (e.g., events may have included demonstrations of "transaction auditing" where changes are logged and displayed in a different grid view) without being told that this feature demands additional coding, often request more hardware to accommodate extra database access.

Users might believe they can demand auditing in every field, whereas developers might think this is *feature creep* because assumptions have been made about the extent of user requirements. If the developer has committed delivery before the user requirements were reviewed, developers are placed between a rock and a hard place, particularly if the user management can achieve an advantage from the failure to implement requirements.

(4) ***Developer attachment to prototype*** — Developers can also become attached to prototypes they have spent a great deal of effort producing; this can lead to problems like attempting to convert a limited prototype into a final system when it does not contain an appropriate underlying architecture. (This may suggest that throwaway prototyping, rather than evolutionary prototyping, should be used.)

(5) ***Excessive development time of the prototype*** — A key property of prototyping is the fact that it is supposed to be performed quickly. If the developers lose sight of this fact, they may try to develop a prototype that is too complex. When the prototype is thrown away the precisely developed requirements that it provides may not yield a sufficient increase in productivity to make up for the time spent developing the prototype. Users can become stuck in debates surrounding details of the prototype, thereby delaying the development team and delaying the final product.

(6) ***Expense of implementing prototyping*** — The start-up costs for building a development team focused on prototyping may be high. Many companies have development methodologies in place, and changing them can mean retraining, retooling or both. Many companies tend to automatically implement the prototyping phases without retraining workers to expected levels of competency.

A common problem with adopting prototyping technology is high expectations for productivity with insufficient effort behind the learning curve. In addition to training for the application of a prototyping technique, there is an often over-looked need for developing corporate and project specific underlying structure to support the technology. When this underlying structure is omitted, the result is often lower productivity [Urban 1992].

5. Best projects to use prototyping.

It has been argued that prototyping, in some form or another, should be consist-ently used. However, prototyping is most beneficial for systems that will en-counter many user interactions.

It has been found that prototyping is very effective in the analysis and design of *on-line systems*, especially for *transaction processing*, where the use of screen dialogs is much more evident. The greater the interaction between the computer and the user, the greater the benefit that can be obtained from building a quick system and letting the user explore it [Crinnion 1991]. Prototyping is especially effective for designing a solid, practical *human-computer interface*.

Systems with little user interaction, such as *batch processing,* or systems that primarily perform calculations benefit little from prototyping. Sometimes, the coding needed to perform the system functions may be too intensive and the potential gains that prototyping could provide are too small.

References

- **[Brooks 1975]** Frederick P. Brooks, *The Mythical Man-Month: Essays on Software Engineering.* Addison-Wesley Publishing Co., Reading, MA, 1975.

- **[Crinnion 1991]** John Crinnion, *Evolutionary Systems Development, A Practical Guide to the Use of Prototyping within a Structured Systems Methodology.* Plenum Press, New York, 1991.

- **[Davis 1991]** Alan M. Davis, Citing: E. Bersoff and A. Davis, "Impacts of Life Cycle Models of Software Configuration Management," *Com. ACM,* Aug. 1991.

- **[Davis 1992]** Alan M. Davis, "Operational Prototyping: A New Develop-ment Approach," *IEEE Software,* September 1992.

- **[Grimm 1998]** Todd Grimm, "The Human Condition: A Justification for Rapid Prototyping," *Time Compression Technologies,* Vol. 3, no. 3. Accel-erated Technologies, Inc., May 1998.

- **[Nielson 1993]** Jacob Nielson, *Usability Engineering*, Academic Press, Cambridge, MA, 1993.

- **[SPC 1997]** "Evolutionary Rapid Development," SPC Document – SPC-97057-CMC, Version 01.00.04, Software Productivity Consortium, Herndon, VA, June 1997.

- **[Urban 1992]** Joseph E. Urban, "Software Prototyping and Requirements Engineering," Rome Laboratory, Rome, NY, 1992.

Chapter 5
A Document Standard for a
Concept of Operations Document[5]

Abstract

A ConOps should state the purpose and scope of the document, reference materials for the document, and definitions and acronyms used within the document. This standard is provided to aid the reader in understanding the ConOps document.

1. Introduction

A ConOps should state the purpose and scope of the document, reference materials for the document, and definitions and acronyms used within the document. This section is provided to aid the reader in understanding the ConOps document.

1.1 Purpose of the ConOps.

This subsection should:

- Identify the document. For example, the ConOps should start with ". . . (insert your organization name, or perhaps individual names in an academic project), propose to develop and deliver to . . . (insert name of customer) a computer program and user manual identified as . . . (insert name of the software system)."

- Briefly describe the specific purpose of the particular ConOps, including why it is being written. Explain what the document is intended to accomplish.

1.2 Scope of the ConOps. This subsection should:

- Identify the document as the ConOps for a particular software product. Describe what the ConOps is or is not supposed to do; for example, 'Will this document be considered a binding contract?' 'Is it in response to a

---

5. This article is modeled after *IEEE Standard 1362-1998*, "IEEE Guide for Information Technology — System Information — Concept of Operation (ConOps) Document"; however, it is not a *direct copy*. The purpose of this document is to (1) aid the *IEEE Software Requirements Certificate of Proficiency* exam candidate to prepare, take and pass the software requirements exam, (2) provide an example of a software requirements standard to demonstrate the usefulness of an IEEE software engineering standard, and (3) provide a template for university students to use when writing an SRS for a classroom software engineering project.

request for proposal?' 'Are the appendices to be considered a formal part of the proposal?'

- If the ConOps documents a system which encompasses more than the specific product that the developer is proposing to create, explain the differences.

- Briefly describe the problem which this document addresses. Provide a level of detail such that individuals in management positions need to read only this introductory section to determine whether or not the remainder of the document is of concern.

- Indicate the current version, past versions and the planned evolution of the document.

1.3 Definitions, acronyms and abbreviations.

This subsection should provide the definitions of all terms, acronyms, and abbreviations required to properly interpret the ConOps. This information may be provided by reference to one or more appendices in the ConOps or by reference to other documents. Only terms actually used in the document should appear here. Consider the intended audience of the document, determining what should be defined; also consider that a ConOps document is intended for the customer and management as well as technical readers.

1.4 References. This subsection should: [Note-identify groups 1 and 2 below]

- Group 1: Provide a list of all documents specifically cited in the document. Only cited references should be included.

- Group 2: Compile a list of additional references to provide background information.

- Identify each document by title, report number if applicable, date, and publishing organization.

- Specify the sources from which the references can be obtained if there is likely to be difficulty in locating the documents. This information may be provided by reference to an appendix or to another document.

1.5 Overview. This subsection should:

- Describe what the remainder of the document contains. It is not necessary to describe Section 1 in this overview, since this is the end of Section 1. We can presume the reader already knows what is included.

- Explain how the body of the document and appendices are organized.

2. The Current System or Situation

This section shall describe the system or situation as it currently exists. If there is no current system on which to base changes, this section describes the situation that motivates development of the proposed system. In this case, the following subsections will be used as appropriate to describe the motivating situation.

This section can be covered by providing a reference to the ConOps document for the existing system, provided that the document accurately reflects the current status of the existing system. In cases where there is no existing system to be replaced or modified, this section can be covered by a needs analysis document for the situation that motivates a new system, provided such documentation exists and provided it adequately addresses the interest of this section.

2.1 Background, objectives and scope.

This subsection shall provide an overview including background and context of the current system or situation. This subsection should also provide an overview of the problem domain. Overview of the contents helps readers to better understand deficiencies in the current system or situation and reasons for the desired changes and improvements.

2.2 Operational policies and constraints.

This subsection shall describe operational considerations of the current system or the current design situation, including operational policies and constraints. Operational policies are predetermined management decisions regarding operation of the current system. Operational policies normally take the form of general statements or understandings that guide decision making activities. For example, an operational policy might require at least two operators in the control room at all times. Operational constraints might include a constraint on the hours of operation of the system, perhaps limited by access to secure terminals.

2.3 Description of the current system or situation.

This subsection shall describe the current system and identify the facilities, equipment, computing hardware, software, personnel, and operational procedures for the existing system. This description should be as detailed as necessary to give readers an understanding of the numbers, versions, capacities, etc., of the operational equipment being used. For example, if the current system contains a database, the capacity of the storage units(s) should be specified, provided that information influences the users' operational capabilities.

Those aspects of safety, security and privacy that exert influence on the operation or operational environment of the current system should be described.

It is important that the description of the current system or situation be simple and clear enough that all intended readers of the document can fully understand it. It is important to keep in mind that the ConOps must be written in the user's language. In most cases, this means avoidance of terminology specific to computers.

3. Justification of Change

This section shall describe the needs that motivate development of a new system or modification of an existing system. It provides a transition from the description of the current system or situation to a description of the proposed system. If there is no current system on which to base changes, this paragraph should provide the justification for a new system.

3.1 Justification of changes and new features.

This subsection shall:

- Describe the new or changed elements of user needs, the mission, objectives, environment, interfaces, personnel, or other factors that require a new system or modification of an existing system.

- Summarize the deficiencies or limitations of the current system or situation that makes it unable to respond to the new or modified factors.

- Provide justification for a new or modified system. If the proposed system is to meet a new threat, describe the reasons why a new system must be developed to meet this threat; if the proposed system improves a current operation, describe the rationale behind the decision to modify the existing system; if the proposed system implements a new functional capability, explain why this function is necessary.

3.1.1 Description of needed changes and new features. This subsection shall summarize new or modified capabilities, functions, processes, interfaces, and other changes needed to respond to the factors identified in Paragraph 3.2. Descriptions of changes to an existing system should be based on descriptions of the current system provided in Paragraph 3.2. If there is no existing system on which changes are to be based, this subsection should summarize the capabilities to be provided by a new system. Needed changes and new capabilities should be described in the users' terminology and from the users' point of view. This description should include, as appropriate:

- *Capability changes* — Description of the functions and features to be added, deleted and modified in order for the new or modified system to meet its objectives and requirements.

- *System processing changes* — Description of the changes in the process(es) of transforming data that will result in new output with the same data, the same output with new data or both output situations.

- *Interface changes* — Description of changes in the system that will cause changes in the interfaces, and changes in the interfaces that will cause changes in the system.

- **Personnel changes** — Description of changes in personnel caused by new system requirements, changes in user classes or both cases.

- **Environmental changes** — Description of changes in the operational environment that will cause changes in the system functions, processes, interfaces, or personnel, and/or changes that must be made in the environment because of changes to the system functions, processes, interfaces, or personnel.

- **Operational changes** — Description of changes to the users' operational policies, procedures, methods, or daily work routines caused by the above changes.

- **Support changes** — Description of changes to the support requirements caused by changes in the system functions, processes, interfaces, or personnel, and/or changes in the system functions, processes, interfaces, or personnel caused by changes in the support environment.

- **Other changes** — Description of other changes that will impact the users, but which do not fit in any of the above categories.

3.1.2 Priorities among changes and new features. This subsection shall identify priorities among the needed changes and new features. Each new or changed feature should be categorized as essential, desirable or optional. Desirable and optional features should be prioritized within their categories.

- **Essential features** — Identification of the essential features that must be provided by the new or modified system.

- **Desirable features** — Identification of the desirable features that should be provided by the new or modified system. Desirable features must be prioritized.

- **Optional features** — Identification of the optional features that might be provided by the new or modified system. Optional features must be prioritized.

Categorizing the desired features as essential, desirable or optional forces users to think more deeply about their true needs than they might otherwise. Categorization of desired features also provides guidance to the system development team during the design of the proposed system. This information is also helpful in the case of budget cuts or schedule overruns, once the priorities among the system requirements are provided. This would allow the low-level requirements and features to be delayed and/or omitted.

3.1.3 Changes and new features considered but not included. This subsection shall identify changes and new features considered but not included in Paragraph 3.2, and the rationale for not including them. By describing changes and features considered but not included in the proposed system, the authors document the results of their analytical activities. This information can be useful to other personnel involved with the system development, whether it be users, developers or the customer, should they want to know if a certain change or feature was considered, and if so, why it was not included.

3.1.4 Assumptions and constraints. This subsection shall describe assumptions or constraints applicable to changes in existing features and the addition of new features scheduled to be added. This should include all assumptions and constraints that will affect users during development and operation of the new or modified system.

3.2 Concept of operations (ConOps) for the proposed system.

This section shall describe the operational features of the proposed system that will result from the desired changes. This section describes the proposed system in a high-level manner, indicating the operational features that are to be provided without specifying design details of the proposed system. Methods of the description to be used and the level of detail included in the description will depend on the situation.

The level of detail should be sufficient to fully explain how the proposed system will fulfill the users' needs and customers' expectations. The method of description should be easily understood by all intended readers of the ConOps document.

3.2.1 Background, objectives and scope. This subsection shall provide an overview of the proposed system including, as applicable, background, mission, objectives, and scope of the system. In addition to providing the background of the proposed system, this subsection should provide a brief summary of the motivation for the system development. The goals for the new or modified system should also be defined, together with the strategies, tactics, methods, and techniques proposed to accomplish them. The modes of operation, classes of users, and interfaces to the operational environment should be summarized in this subsection and defined in detail in subsequent sections.

3.2.2 Operational policies and constraints. This subsection shall describe operational policies and operational constraints for the proposed system. An operational policy might state that operational personnel should have a specified amount of training and experience to perform certain system functions.

3.2.3 Description of the proposed system. This subsection shall provide a description of the proposed system, including as appropriate:

- The operational environment and its characteristics.

- Major hardware, software and manual system components, including databases and the interconnections among these components.

- Interfaces to external systems, software or procedures.

- Capabilities and operational features of the proposed system.

- Overview of inputs, outputs, data flow, and manual and automated processes sufficient to understand the proposed system from the user's point of view.

- Performance characteristics such as speed, throughput, volume, and frequency of operations that impact the users' view of the proposed system.

- Quality attributes such as reliability, maintainability, availability, flexibility, portability, usability, and efficiency that are visible to users.

- Provisions for safety, security, privacy, and continuity of operations in the event of emergencies.

This subsection contains a major portion of the description of the proposed system. Graphical representations should be used wherever possible, especially since ConOps documents should be understandable by several types of readers.

3.2.4 Modes of operation for the proposed system. This subsection shall describe the various modes of operation for the proposed system, which might include normal, degraded, maintenance, training, emergency, alternate-site, active, and/or idle modes. Important modes to include are backup and emergency modes, if they exist. This is especially true if these modes involve different geographical sites and equipment that have significant impacts on the system. This subsection can be organized into subparagraphs, one for each mode described. System processes, procedures, capabilities, and functions should be described for each mode.

3.2.5 User classes for the proposed system. This subsection shall distinguish the different user classes of the proposed system. A user class is distinguished by the patterns of interaction between user and system. Factors that distinguish a user class include responsibilities, skill level, work activities, and system modes exercised by the user. Different user classes may have distinct operational scenarios for their interactions with the system. In this context, a user is anyone who will interact with the proposed system, including operational users, data entry personnel, system operators, operational support personnel, software maintainers, and trainers.

3.2.6 Organizational structure. This subsection shall describe the organizational structure of the various user groups and user classes that will be involved with the proposed system. Organizational charts are useful graphical tools for this purpose.

(1) **Profiles of user classes** — This subsection shall provide a profile of each user class for the proposed system. If some users play several roles, each role should be identified as a separate user class.

(2) **Interactions among classes** — This subsection shall describe interactions among the various classes of users that will be involved with the proposed system. In particular, interaction among user groups, operators and maintainers should be described. Interactions that will occur among the users of the proposed system, and between users and non-users, both within the organization and across interfacing organizations, should be described if they are relevant to the operation of the proposed system.

3.2.7 Other involved personnel. This subsection shall describe other personnel who will not directly interact with the proposed system, but who will have an influence on, and be influenced by, the proposed system.

3.2.8 Support environment for the proposed system. This subsection shall describe the support concepts and support environment for the proposed system, including the support agency; facilities; equipment; support software; repair and replacement criteria; maintenance levels and cycles; and storage, distribution, and supply methods.

3.3 Operational scenarios for the proposed system.

This section shall provide scenarios of operation for the proposed system by user class and mode of operation. A scenario is a step-by-step description explaining how users will interact with the proposed system under various circumstances. Scenarios should be described in a manner that will allow readers to walk through them and gain a detailed understanding of the proposed system.

Scenarios play several important roles. The first is to bind together, into a comprehensible whole, all of the individual pieces of information regarding the system. Scenarios help the readers of a ConOps document to understand how the system will provide the operational capabilities. In addition, scenarios can serve as the basis for a first draft of the users' manual, and as the basis for developing acceptance test scenarios. The operational scenarios also provide a mechanism for verifying that the system design satisfies the users' needs and customer's expectations.

Scenarios are an important component of a ConOps, and should therefore receive substantial emphases. The number of scenarios and level of detail specified should be proportional to the perceived risk of insufficient communication among the users, customer and developer.

3.4 Summary of impacts.

This section shall describe the operational impacts of the proposed system on users, customers, developers, and the operational support and maintenance organizations during development and operation of the envisioned system.

This information is provided in order to allow all affected organizations to pre-pare for changes that will be brought about by the new system and to allow for planning of the impacts on those organizations during the development of, and change to, the new system.

3.4.1 Operational impacts. This subsection shall describe the anticipated oper-ational impacts on the user, development, support, and maintenance agency during operation of the proposed system. These impacts may include interfaces with primary or alternate computer operating centers; changes in procedures; use of new data sources; changes in quantity, type and timing of data to be input into the system; changes in data retention requirements; new modes of opera-tion based on emergency, disaster or accident conditions; and new methods for providing input data if the required data is not readily available.

3.4.2 Organizational impacts. This subsection shall describe the anticipated organizational impacts on the user, development, support, and maintenance agency during operation of the proposed system. These impacts may include modification of responsibilities; addition or elimination of job positions; training or retraining of users; changes in numbers, skill levels, position identifiers, or locations of personnel; and numbers and skill levels of personnel needed for contingency operation at one or more alternate sites following an emergency, disaster or accidents.

3.4.3 Impacts during development. This subsection shall describe the antici-pated impacts on the user, development, support, and maintenance agency dur-ing the development project for the proposed system. These impacts may in-clude involvement in studies, meetings and discussions prior to award of con-tract; user and support involvement in reviews and demonstrations; evaluation of initial operating capabilities and evolving versions of the system; develop-ment or modification of databases and required training; parallel operation of the new and existing systems; and operational impacts during system testing of the proposed system.

3.5 Analysis of the proposed system.

This section shall provide an analysis of the benefits, limitations, advantages, disadvantages, and alternatives and trade-offs considered for the proposed sys-tem.

3.5.1 Summary of improvements. This subsection shall provide a qualitative summary of the benefits to be provided by the proposed system. This summary should include the following, as applicable. In each case, the benefits should be related to deficiencies of the existing system or situation identified.

- *New capabilities* — New features or functionality.

- *Enhanced capabilities* — Upgrades to existing capabilities.

- *Deleted capabilities* — Unused, confusing or dangerous capabilities removed.

- *Improved performance* — Better response time, reduced storage requirements, improved quality, etc.

3.5.2 Disadvantages and limitations. This subsection should provide a qualitative summary of the disadvantages and/or limitations of the proposed system. Disadvantages might include the need to retrain personnel, to rearrange work spaces, and to learn a new style of user interface. Limitations might include features desired by users but not included in the delivered system, and to degradation of existing capabilities to gain new capabilities or greater than desired response times for certain complex operations.

3.5.3 Alternatives and trade-offs considered. This subsection should describe the features considered for inclusion in the proposed system, the trade-offs among them and rationales for the decisions made. In the context of a ConOps, alternatives are operational alternatives and not design alternatives except to the extent that the operational capabilities desired in the new system may be limited by the state of technology, user and/or customer imposed design constraints and the resulting design alternatives. This information can be useful to determine, now and at later times, whether a given capability was considered, or why a particular feature was rejected. This information would probably be lost if not included here.

Chapter 6
A Document Standard for a
Software Requirements Specification[6]

Abstract

Engineering standards are documented agreements containing technical specifications or other precise criteria to be used consistently as engineering processes, rules, guidelines, and definitions of characteristics, to ensure that materials, products and services are fit for their purpose.

Standards provide rules, guidelines and heuristics which, if followed, deliver an assurance of "good practice"—they are not intended to be about "best practice."

The result of using this standard is a complete and reasonably correct software requirements specification (SRS). The SRS is a description of a software system to be developed. It lays out functional and non-functional requirements, and may include a set of use cases that describe user interactions that the software must provide.

The software requirements specification establishes the basis for an agreement between customers and contractors or suppliers (in market-driven projects, these roles may be played by the marketing and development divisions) detailing what the software product is expected to do as well as what it is not expected to do. Software requirements specifications permit a rigorous assessment of requirements before design can begin and reduce later redesign. They should also provide a realistic basis for estimating product costs, risks and schedules.

1. Introduction to the Software Requirements Specifications

This section of the SRS should state the purpose and scope of the SRS, the planned evolution of the SRS, reference materials for the SRS, and definitions and acronyms used within the SRS.

1.1 Purpose of the SRS.

This subsection should determine the following:

(1) ***Specific purpose*** — Delineate the special purpose of the particular SRS.

---

6. This article is modeled after *IEEE Standard 830-1998*, "IEEE Recommended Practice for Software Requirements"; however, it is <u>not</u> a *direct copy*. The purpose of this document is to (1) aid the IEEE *Software Requirements Certificate of Proficiency* example of a software requirements standard to demonstrate the usefulness of an IEEE software engineering standard, and (3) provide a template for university students to use when writing an SRS for a classroom software engineering project.

(2) *Audience* — Specify the *intended audience* for the SRS.

1.2 Scope of the SRS.

This subsection should:

- *Identify* —The software products to be produced by name.

- *Explain* —What the software product will and, if necessary, will not do.

- *Indicate* —The current version, past versions and the planned evolution of the SRS.

1.3 Definitions, acronyms and abbreviations.

This subsection should provide the definition of all terms, acronyms and abbreviations required to properly interpret the SRS. This information may be provided by reference to one or more appendices in the SRS or by reference to other documents.

Only terms actually used in the SRS should be defined here. This is *not* a dictionary of software engineering terms.

1.4 References.

This subsection should:

(1) *Provide* — A complete list of all documents referenced elsewhere in the SRS.

(2) *Identify* — Each document by title; report number, if applicable; data; and publishing organization.

(3) *Specify* — The sources from which the references can be obtained.

This information may be provided by reference to an appendix or to another document.

1.5 Overview.

This subsection should:

(1) *Describe* what the remainder of the SRS contains. It is not necessary to describe Section 1 in this overview, since this is the end of Section 1. We can presume the reader already knows what is included.

(2) *Explain* how the SRS is organized. Be sure to include an explanation of the appendices if they are considered to be a formal part of the SRS.

2. General Description

This section of the SRS should describe the *general factors* that affect the product and its requirements. This section does not state specific requirements; it only provides information that makes the requirements stated in Paragraph 3 (Specific Requirements) easier to understand.

This section usually consists of five subsections as follows:

1. Product perspective.
2. Product functions.
3. User characteristics.
4. General constraints.
5. Assumptions and dependencies.

2.1 Product perspective.

This subsection of the SRS puts the product into perspective with other related products or projects.

(1) If the product is *independent* and *self-contained*, this should be stated here.

(2) If the SRS defines a product that is a component of a larger system or project, as frequently occurs, then this subsection should:

   a. *Describe* the functions of each component of the larger system or project and identify their interfaces.

   b. *Identify* the principal external interfaces of this software product. Note: This is not a detailed description of these interfaces; the detailed description is provided elsewhere in the SRS.

   c. *Describe* the computer hardware and peripheral equipment to be used. Note: This is an overview description only.

A block diagram showing the major components of the larger system or project, their interconnections and their external interfaces can be very helpful.

This subsection should not be used to impose a specific design solution or specific design constraints affecting the solution. This subsection should provide the reasons why certain design constraints are later specified as part of the Specific Requirements Section of the SRS.

2.2 Product functions.

This subsection of the SRS should provide a *summary* of the functions that the software will perform.

For example, an SRS for an accounting program might use this subsection to address customer account maintenance, or customer statement and invoice preparation without mentioning the vast amount of detail each of these functions requires.

Sometimes the *function summary* necessary for this part can be taken directly from the section of the higher-level specifications (if one exists) that allocates particular functions to the software product.

Note that for the sake of clarity:

- The *functions* should be organized so as to make the list of functions understandable to the customer or to first time readers.

- Block diagrams *showing the different functions* and their relationships can be helpful. Remember, however, that such a diagram is not a requirement for the design of a product itself; it is simply an effective explanatory tool.

This subsection should not be used to state specific requirements. It should provide the reasons why certain specific requirements are later specified as part of the Specified Requirements Section of the SRS.

2.3 User characteristics.

This subsection of the SRS should describe those general characteristics pertaining to the *eventual users* of the product that will affect the specific requirements.

Many types of people interact with a system during the Operation and Maintenance Phase of the software life cycle. Some of these are users, operators. and maintenance and systems personnel. Certain characteristics of these people, such as educational level, and experience and technical expertise, impose important constraints on the system's operating environment.

For example, if most users of the system are occasional users, a resulting specific requirement might be that the system contain reminders to perform essential functions, rather than assuming that the user will remember these details from the last session or from reading the user's manual.

This subsection should not be used to state specific requirements or to impose specific design constraints on the solutions. This subsection should provide the reasons why certain specific requirements or design constraints are later specified as part of the Specific Requirements Section of the SRS.

2.4 General constraints.

This subsection of the SRS should provide a general description of any other items that will limit the developer's options for designing the system. These can include:

- Regulatory policies.

- Hardware limitations (signal timing requirements, memory size).

- Interfaces to other applications.

- Parallel operations.

- Command requirements (menu driven).

- Audit functions.

- Control functions.

- Higher-order language requirements.

- Signal handshake protocols.

- Criticality of the application.

- Safety and security considerations.

This subsection should not be used to impose *specific requirements* or *specific design constraints* on the solution. This subsection should provide the reasons why certain specific requirements or design constraints are later specified as part of the Specific Requirements Section of the SRS.

2.5 Assumptions and dependencies.

This subsection of the SRS should list each of the factors that affect the requirements stated in the SRS. An *assumption* is something that is taken to be true because we are either unable or unwilling to determine the actual truth. A *dependency* is a condition, environment or activity that is necessary for the software to be built according to these specifications. Any changes to the assumptions or dependencies can affect the requirements stated in the SRS. For example, an assumption might state that a specific operating system will be available on the hardware designated for the software product. If, in fact, the operating system is not available, the SRS would then have to change accordingly.

3. Specific Requirements

This section of the SRS should contain all the details needed by the software developer to create a design. This is typically the largest and most important part of the SRS.

(1)   External interface requirements.

(2)   Functional requirements.

(3)   Performance requirements.

(4)   Design constraints.

(5)   Quality attributes.

3.1 External interface requirements.

This subsection of the SRS describes user interfaces, hardware interfaces, software interfaces, and communication interfaces.

3.1.1 User interfaces. This subsection should specify the characteristics that the software must support for each individual human interface with the software product. For example, if the user of the system operates through a display terminal, the following should be specified:

(1)   Required screen formats.

(2)   Page layout and content of any reports or menus.

(3)   Relative timing of inputs and outputs.

(4)   Availability of some form of programmable function keys.

(5)   The use of menus or command language.

This may simply comprise a list of factors that illustrate how the system will appear to the user. One example might be a requirement for the option of long or short error messages. Like all others, these requirements should be verifiable, saying, for example, "a clerk typist Grade 4 can do function X in Z minutes after one hour of training" rather than "a typist can do function X". (Note: This factor might also be specified in the 'Attributes' section under a subsection entitled "Ease of Use").

3.1.2 Hardware interfaces. This subsection should specify the logical characteristics of each individual interface between the software product and the hardware components of the system. It also covers such topics as devices to be supported, how they are to be supported and protocols.

3.1.3 Software interfaces. This subsection should specify the use of other required software products (for example, a database management system, an operating system or a mathematical package) and interfaces with other application systems (for example, the linkage between an accounts receivable system and a general ledger system).

There are two different types of information which must be provided for software interfaces.

(1) For each required software product, the following should be provided:

   a.   Name.

   b.   Mnemonic.

   c.   Specification number.

   d.   Version number.

   e.   Source.

(2) For each individual interface, this part should:

   a.   Discuss the purpose of the interfacing software as related to this software product.

b. Define the interface in terms of message content and format. (It is not necessary to detail any well-documented interface, but a reference to the document defining the interface is required.)

3.1.4 Communications interfaces. This subsection should specify the various interfaces to communications such as local network protocols, etc.

3.2 Functional requirements.

(See Table 1 when using a functional hierarchy and Table 2 when using an objects hierarchy.)

Table 1:SRS functional requirements (structured format)

Functional requirements. This subsection of the SRS should specify how the inputs to the software product should be transformed into outputs. It describes the fundamental actions that must take place in the software. For each class of functions, or sometimes for each individual function, it is necessary to specify requirements on inputs, processing and outputs. These are usually organized using the following four subparagraphs.

Each function shall be identified as a separate entity, i.e., 1 through n.

(1) **Introduction** — This subsection should provide a description of the purpose of the function and the approaches and techniques employed. It should contain any introductory or background material which might clarify the intent of the function.

(2) **Inputs** — This subsection should contain a detailed description of all data input to this function to include:

   a. Sources of the inputs.

   b. Quantities.

   c. Units of measure.

   d. Timing.

   e. Ranges of the valid inputs to include accuracies and tolerances.

If a data dictionary is included with the SRS, then this section need only name the data items. The descriptions may be found in the data dictionary.

(3) **Processing** — This subsection should define all of the operations to be performed on the input data and the intermediate parameters to obtain the output. It includes specifications of:

   a. Validity checks on the input data.

   b. The exact sequence of operations to include timing of events.

   c. Responses to abnormal situations.

     d.   Any methods (equations, mathematical algorithms and logical operations) which must be used to transform the system inputs into their corresponding outputs.

     e.   Validity checks on the output data.

(4) **Outputs** — This subsection should contain:

     a.   A detailed description of all data output from this function.

     b.   As with inputs, much of this information may be found in the data dictionary. If a data dictionary is included as an appendix to the SRS, it is not necessary to repeat the information. All inputs and outputs must appear in the data dictionary.

     c.   References to interactive specifications or interface control documents where appropriate.

In addition, for those systems whose requirements focus on input/output behavior, the SRS should specify all of the significant input/output pairs and sequences of pairs.

Table 2: SRS functional requirements (object-oriented format)

Functional requirements.

Classes/Objects. This subsection shall contain all class/object information. This information consists of descriptions, attributes, functions, and messages. Each class or object shall be identified as a separate entity, i.e., 1 thru n.

(1) **Class/Object description.** This subsection shall describe for each class/object the class/object by name and a brief explanation of what it represents.

(2) **Attributes (of the SRS).** This subsection shall describe the attributes associated with the class/object. This shall include all attributes, both direct and inherited.

(3) **Functions (of the SRS).** This subsection shall describe all functions associated with the class/object. This description includes all services and methods, both direct and inherited.

3.3 Performance requirements.

Messages (of the SRS). This subsection shall describe all messages associated with the class/object. These messages may be either received or

This subsection should specify both the static and the dynamic numerical requirements placed on the software or on human interaction with the software, as a whole.

(1) *Static numerical requirements* may include:

a. The number of terminals to be supported.

b. The number of simultaneous users to be supported.

c. The number of files and records to be handled.

d. The sizes of tables and files.

(2) *Dynamic numerical requirements* may include, for example:

a. The number of transactions and tasks.

b. The amount of data to be processed within certain time periods for both normal and peak workload conditions.

All of these requirements should be stated in measurable terms, for example, "95% of the transactions shall be processed in less than one second", rather than, "operator shall not have to wait for the transaction to complete".

3.4 Design constraints.

This subsection shall specify the design constraints that can be imposed by other standards, hardware limitations, etc.

(1) ***Standards compliance*** — This subsection should specify the requirements derived from existing standards or regulations.

(2) ***Hardware limitations*** — This subsection should include requirements for the software to operate inside various hardware constraints. For example, these could include:

a. Hardware configuration characteristics (number of ports, instruction sets, etc.).

b. Limits on primary and secondary memory.

3.5 Quality attributes.

This subsection shall describe attributes that can place specific requirements on the software. Some of these are indicated below. These should not be considered as a complete list.

- Maintainability.
- Reliability.
- Safety.
- Security.
- Survivability.
- User friendliness.

3.6 Other Requirements.

This subsection shall contain certain requirements that may, due to the nature of the software or the user organization, be placed in separate categories as indicated below.

3.6.1 Database. This subsection should specify the requirements for any database that is to be developed as part of the product. This might include:

- The types of information identified in user interfaces.

- Frequency of use.

- Accessing capabilities.

- Data element and file descriptors.

- Relationship of data elements, records and files.

- Static and dynamic organization.

- Retention requirements for data.

Note: If an existing database package is to be used, this package should be named under interfaces to software and the details for using it should be specified here.

3.6.2 Operations. This subsection should specify the normal and special operations required by the user such as:

- The various modes of operations in the user organization.

- Periods of interactive operation and periods of unattended operations.

- Data processing support functions.

- Backup and recovery operations.

Note: This is sometimes specified as part of the user interfaces section.

3.6.3 Site adaptation. This subsection should:

- Define the requirements for any data or initialization sequences that are specific to a given site, mission or operational mode.

- Specify the site or mission-related features that should be modified to adapt the software to a particular installation.

Chapter 7
Software Requirements Exercises

These exercises are provided to encourage you to browse the chapter looking for answers to the questions provided. If the truth be told, the correct answer for all software engineering questions is "it depends." To avoid this issue, a set of possible answers are provided. There is (supposedly) only one correct answer.

If you are using this book as a textbook in a university course, your instructor may require you to justify your answer. Why are some of the possible answers correct, while others are wrong? The instructor might also ask you to explain the assumption depended upon when deciding your answer.

However, if you are very clever, maybe you can come up with more than one correct answer (which of course must be justified).

1. **Which of the following is a good reason for using a rapid prototype?**

    [a] A prototype can be used as a basis for the final system delivery.
    [b] Response times and other performance issues can be verified.
    [c] Prototypes are expensive. Therefore, they can be useful as a backup system for the customer.
    [d] A prototype is constructed to display key functionality—even though many features such as error handling may be missing.

2. **Concerned about requirements tracing, a project manager is deciding how to perform this task correctly. She started by compiling the following list of possible attributes of the tracing activity. Which one is correct?**

    [a] Tracing is done forward only
    [b] Tracing includes design and coding only
    [c] A changed requirement can be modified without tracing
    [d] Tracing is performed both forward and backward

3. **The person responsible for managing the requirements is which of the following?**

    [a] The chief system engineer
    [b] The project manager
    [c] The technical manager
    [d] The customer

4. Which of the software life cycles (processes) requires ALL the requirements specifications to be completed prior to beginning design and implementation?

I. Evolutionary development model
II. Incremental development model
III. Spiral model
IV. Hacking model
V. Baseline management (waterfall) model

[a] I and III
[b] II and V
[c] I and II
[d] I, II, and III

5. Which of the following activities would probably NOT be a problem in determining software requirements?

I. Conflict of information supplied by different customers
II. Difficulty in determining the requirements
III. Using the software requirements as a basis of software project costs
IV. Customers are well versed in the requirements of the system
V. Requirements that change after the project starts
VI. Design that changes without the corresponding change in software requirements

[a] I, II, V, and VI
[b] III and IV
[c] III, IV, and V
[d] V and VI

6. A program support librarian performs which of the following functions?

[a] Performs project control
[b] Acts as chief programmer
[c] Maintains the official copy of the project documentation during the development phase
[d] Reviews and approves all new changes to the system

7. Software requirements verification includes which of the follow-ing?

I. Assurance that software requirements comply with system re-quirements
II. Assurance that the software engineer has understood the re-quirements
III. Assurance that the requirements documents conform to com-pany standards
IV. Assurance that the requirements document is understandable, consistent and complete

[a]   I, II and III
[b]   I, II and IV
[c]   I, III and IV
[d]   I, II, III and IV

8. Which of the following non-functional requirements is the most dif-ficult to quantify?

[a] Testability
[b] Understandability
[c] Modifiability
[d] Performance

9. A ConOps document describes which of the following?
[a] A proposed system from the users' point of view
[b] A proposed system from the developer's point of view
[c] An existing system from the users' point of view
[d] An existing system from the developer's point of view

10. If you are assigned a project with a fixed requirements and sched-ule, what is the third parameter that must be flexible?

[a] Cost
[b] Opportunity
[c] Risk
[d] Quality

11. A software system has to be 95% reliable for a particular usage. This is an example of which type of requirement?

[a] A product requirement that is directly measurable
[b] A process requirement that is directly measurable
[c] A product requirement that is NOT directly measurable
[d] A process requirement that is NOT directly measurable

12. Which of the following requirements would you consider being dynamic performance requirements?

I. 25 terminals should be supported
II. 90% of transactions should be processed within two seconds
III. Less than 5% of code should be host dependent
IV. Functions A and B should be allocated to different modules

[a] I only
[b] II only
[c] II and III
[d] II and IV

13. A software requirements specification would normally be expected to contain which two of the following types of requirements?

I. Project requirements
II. Process requirements
III. Functional requirements
IV. External interface requirements

[a] I and II
[b] I, III and IV
[c] I and IV
[d] II and III

14. Which is the characteristic of a good software requirements specification?

[a] Complete
[b] Hierarchical
[c] Verifiable
[d] Traceable

15. Which of the following are useful in gathering software requirements information?

    I.   Interviews
    II. Prototypes
    III. Scenarios
    IV. Statements of work
    V.   Feasibly studies

    [a] I, II, and V
    [b] I, III, IV and V
    [c] I, II, IV
    [d] I, II, III

16. Which of the following are INCORRECT statements about software requirements activities?

    [a] Software requirements analysis is the process of determining what
        will be delivered at the end of the software project.
    [b] Software requirements specifications are the descriptions of the
        software requirements.
    [c] Software design is the act of determining how the software will be
        built to meet the software requirements specifications.
    [d] Software requirements validation is the process of determining if the
        software requirements properly implemented the system design al-
        located to software.

17. Which of the following illustrates the notion of portability?

    [a] Moving a software system from one computer system to another with
        a minimum of modifications
    [b] Copying of a design on a new system
    [c] Taking an existing module or code fragment and using it to create a new
        product
    [d] Using (part of) a manual when writing the documentation for a new
        product

18. Which of the following non-functional requirements is difficult to quantify?

    [a] Testability
    [b] Understandability
    [c] Modifiability
    [d] Performance

19.   **A ConOps document describes which of the following?**

[a]   A proposed system from the users' point of view
[b]   A proposed system from the developer's point of view
[c]   An existing system from the users' point of view
[d]   An existing system from the developer's point of view

20.   **What document should be produced during the software require-ments phase?**

I.   **Concept of operations document (ConOps)**
II.   **Draft users' manual**
III.   **Project plan**
IV.   **Architectural design description**
V.   **Software interface requirements specification**
VI.   **Software requirements specification**
VII.**Test plans**

[a]   I, III, and VI
[b]   I, IV, V, and VI
[c]   II, Ill, V, VI, and VII
[d]   II, V, VI, and VI

INDEX

joint application development (JAD) meeting, 17

major issues of software engineering requirements, 4, 6
market analysts, 15
Mellor, S.J., 22
mini-specs, 22
miscellaneous requirements, 10
model validation, 4, 40
modes of software engineering requirements, 4, 7

Nielsen, Jacob, 96
nonfunctional requirements, 19

object model, 24, 57
observation, 18, 55
operational requirements, 19

performance requirements, 9, 10, 30, 109, 111
Pirbhai, I.A., 22
primary requirements, 20
priority levels, 20
process actors, 4, 14
process requirements, 9, 19, 21, 118
product requirements, 8, 19
project management responsibilities, 4, 40
project manager, 2, 14, 28, 40, 115
project plan, 2
prototypes, vi, 17, 39, 88, 89, 115
PSL/PSA tool, 23

quality attribute, 9. 10, 11, 28, 30
quantifiable requirements, 4, 11

regulators, 15
requirements allocation, 25
requirements analysis, 4, 18-19, 41, 55, 62-63, 66
requirements attributes, 4, 38
requirements documents, 2
requirements elicitation, 15, 51
requirements management tools, 41
requirements management, v, 4, 40, 60, 61, 65
requirements modeling tools, 4, 21, 41
requirements negotiation, 4, 26
requirements reviews, 4, 37, 64

requirements sources, 4, 16, 52
requirements specifications, v, vi, 4, 26, 49, 61, 105
requirements verification, validation and testing, v, 35
rework, 13, 29, 80
Ross, Doug, 22
Rumbaugh, James, 24

scenarios, 17, 54, 101, 119
software capability, 6
software developer, 15
software engineering requirements tools, 41
software engineering requirements, 1, 6
software engineering requirements, modes of, 4, 7
software prototyping, 39, 87, 89
software quality attributes, 30
software requirements analysis, 3
software requirements elicitation, 3
software requirements fundamentals, 3
software requirements management, 3
software requirements process model, 12
software requirements processes, 3, 12
software requirements specification, contents of, 4, 30
software requirements specifications, vi, 2, 3, 49, 105
software requirements validation, 3
software technical manager tools, 42
stability, 20, 31, 32
stakeholder, 14, 15-81passim
structured analysis and design technique (SADT), 22
structured analysis, 21, 22
system requirements, 4, 5, 16, 28
system testing, 36

technical management, 40, 41
technical requirements, 8, 19, 27, 28
Teichroew, Daniel 23
testing levels, 4, 36
tracing, 4-115 passim

unit testing, 36
users, 14, 84, 91, 92

validation, v, 35, 36, 37, 61, 63, 64
verification, v, 35, 36, 61
volatility, 20

NOTES

NOTES

NOTES

NOTES

NOTES

www.ingramcontent.com/pod-product-compliance
Lightning Source LLC
Chambersburg PA
CBHW080559220326
41599CB00032B/6536